3 Chemistry of Plant Protection

Pyrethroid Residues, Immunoassays for Low Molecular Weight Compounds

With contributions by W. Blaß and E. W. Weiler

With 10 Figures and 97 Tables

Springer-Verlag
Berlin Heidelberg New York
London Paris Tokyo Hong Kong

Managing Editors:

Dr. G. Haug
Pflanzenschutzzentrum Monheim, Bayerwerk
D-5090 Leverkusen/FRG

Prof. Dr. H. Hoffmann
Leiter des Ressorts Chemische Forschung
Pflanzenschutzzentrum Monheim, Bayerwerk
D-5090 Leverkusen/FRG

This series continues the handbook
"Chemie der Pflanzenschutz- und Schädlingsbekämpfungsmittel"
edited by R. Wegler

ISBN-13:978-3-642-74848-6 e-ISBN-13:978-3-642-74846-2
DOI: 10.1007/978-3-642-74846-2

Softcover reprint of the hardcover 1st edition 1990

2152/3020-543210 — Printed on acid-free paper

Editorial Board

Editorial

The series "Chemistry of Plant Protection" is the successor to our handbook "Chemie der Pflanzenschutz- und Schädlingsbekämpfungsmittel" which was edited by R. Wegler.

Aims and Scope

Plant protection chemistry continues to develop rapidly. Important targets have been achieved and improvements with respect to selectivity, efficacy, mode of action, environmental and toxicological compatibility have been made. There have been new developments in the field of formulation and side-effect-research, and new classes of substances with high effectivity at very low dosages have been discovered.

These developments have been accelerated by new biological and biochemical discoveries as well as by advances in synthetic chemistry. In order to emphasize the interaction between these of different fields of crop protection chemistry, such as insecticide, fungicide or herbicide research, has been abandoned. The following volumes will contain recent developments in research into new active chemical substances as well as reports on metabolism, residue analysis, biochemical mechanisms and other important innovations.

The volumes will be more or less topic-oriented. Each volume will have a "volume index" which approximately characterizes the contents. The aim of the publisher and the board of editors is to produce review articles of the highest quality by leading scientists in the field of plant protection. Suggestions for such contributions from all those involved in plant protection are, of course, always welcome.

Editorial Policy

The series publishes critical review articles in English from invited authors and covering new developments in this field. The topics should be covered comprehensively and the international literature evaluated. Ideally, contributions should be 40–80 typewritten pages long. Experimental details, except when not previously published, should be covered by citing appropriate references. — The responsible editor, that is the editor who invited the article, discusses the scope of the review with the author on the basis of a tentative outline which the author is asked to provide.

Editors Publisher

Table of Contents

Residues of Pyrethroid Insecticides and Their Analytical Determination

Wolfgang Blaß
Bayer AG, 5090 Leverkusen, Bayerwerk, FRG

Dedicated with sincere gratitude to my esteemed academic teacher and doctoral supervisor, Professor Dr. Ludwig Acker, Münster, FRG, on occasion of his 75th birthday.

1 Introduction

The modern synthetic pyrethroids owe their existence to pyrethrin I, the insecticidally most effective component of the pyrethrum extract, which is derived from Chrysanthemum roseum (habitat: Near East) and Chrysanthemum cinerariifolium (habitat: Dalmatia). These plants belong to the composite family (Compositae) and are meanwhile being grown commercially on a large scale especially in Kenya and Tansania.

The structure elucidation of pyrethrin I eventually lead to the synthesis of numerous structurally related compounds, the pyrethroids, particularly since very soon the world demand could no longer be met with the pyrethrum supply of natural origin.

Pyrethrum can be considered an "ideal insecticide": it is effective against a great number of pests without having a too broad spectrum of activity, and it is regarded to be relatively safe to warm-blooded animals [1].

The insecticidally active components of the pyrethrum extract and the first synthetically produced pyrethroids resmethrin, bioresmethrin, allethrin, bioallethrin, S-bioallethrin and cismethrin are relatively instable upon exposure to light and air (oxygen). These properties finally limited the use of the above mentioned products to the areas of household and hygiene, the veterinary field and the prevention of losses or spoilage of stored products due to insect infestations.

Since about 1973, pyrethroids are on the market which are considerably more stable to environmental factors, even though they are, on the other hand, somewhat more toxic to warm-blooded animals.

Meanwhile, numerous pyrethroids form an essential part of the great variety of pest control agents which are effective against insects and mites and are used worldwide in modern agriculture.

There is no doubt that chemical plant protection measures belong to the most important weapons in fighting malnutrition and hunger in the world. This has resulted in an increasing demand for such products within the recent decades.

The increasing use of plant protectants, in turn, has led to a stronger request by the public for "safe" products and a reasonable use of them.

One aspect in the efforts to make plant protectants as safe as possible for the consumer is the investigation of their residual behaviour.

In this monograph, an overview will be presented of residue analytical techniques and residue data reported in the literature. Unpublished results from studies conducted in own laboratories will also be included. It should be pointed out that details about residues of degradation products (metabolites) of pyrethroids are rather seldomly reported in literature. There are mainly two reasons for this:

1. Rather often no maximum residue limits are established for degradation products so that any values having been determined would ultimately not be justiciable.
2. Frequently the unchanged parent compound turns out to be the main component when analysing harvest samples for residues.

As a rule, commercial pyrethroid formulations contain mixtures of diastereomeric, optical and/or geometrical isomers [2]. Since the biological effectiveness depends upon the geometry and chirality of the pyrethroid molecules [3], it is of interest to analytically detect the biologically most active isomer in the respective analytical sample.

The number of published residue analytical methods for pyrethroids has not yet grown too large. E. Papadopoulou-Mourkidou in her article [4] gives an overview of the residue analytical methods published in the international scientific literature until 1982. This covered the pyrethroid insecticides allethrin, cypermethrin, decamethrin, fenvalerate, permethrin, phenothrin, resmethrin and tetramethrin which had been introduced into the market until that time. The author concluded that high performance liquid chromatography (HPLC) and gas chromatography (GLC) are the methods of choice for the trace-analytical determination of pyrethroids. Concerning the latter method it must be added that halogen-free pyrethroids (chrysanthemate type pyrethroids) should be analysed using the flame ionization detector (FID), while for all others the especially halogen-sensitive electron capture detector (ECD) should be used.

The following survey covering analytical methodology will, therefore, mainly deal with the progress made since 1982 and with supplementary information obtained or published since that time.

2 Residue Analysis of Pyrethroids

2.1 Analytical Methods

In Table 1 those pyrethroids are presented for which residue analytical methods or residue data are available. Further information about the chemistry of these substances and their biological properties can be taken from the publications by K. Naumann [2] and [5].

A residue analysis usually involves the following steps:

1. Sample preparation;
2. Extraction;
3. Clean-up of the extract;
4. Qualitative and quantitative determination of the residues;
5. Confirmation of the identity in case of positive findings.

Ref. [1]: The quality of an analytical result is considerably influenced by the kind of sampling. The rules of statistics should be followed when taking samples in order to guarantee that the material to be analysed is representative of the situation prevailing on the treated area or of the sample "drawn from the market". Useful hints concerning sampling can be taken from Chapter VIII of the "Compilation of methods for residue analysis of plant protectants" of the Deutsche Forschungs-Gemeinschaft (DFG) (German Society for the Advancement of Scientific Research), Verlag Chemie, Weinheim, Bergstraße (available in English) or from the "Guide to Codex Recommendations Concerning Pesticide Residues", Part 5, Codex Alimentarius Commission, FAO/WHO (1984).

Ref [2]: Normally, organic solvents are used for the extraction of the compounds from the material to be investigated. High speed blenders such as ®Ultra Turrax (Janke and Kunkel 7813 Staufen/Breisgau, FRG) or ®Polytron (Kinematica, CH 6010 Kriens) are considered as suitable technical aids. A crucial point requiring special attention is the quantitative extraction. This is attempted by using different solvents or solvent systems for one and the same material to be extracted. However, the efficiency of such extraction procedure can only by checked by applying radioacti-

Table 1. Pyrethroid insecticides, formulas and names

Entry No.	Formula	Stereochemistry	No. of Isomers	Common name	Trade name	Basic producer, distributor
1	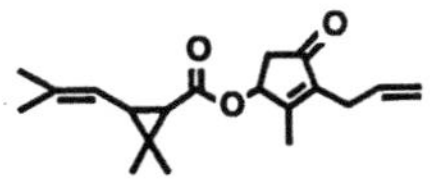	(±) cis/trans; RS	8	Allethrin	Alleviat, Pynamin	Mc Laughlin, Sumitomo
				Pallethrin		Roussel-Uclaf
		1 R trans S		Bioallethrin	Bioallethrin	Procida
2	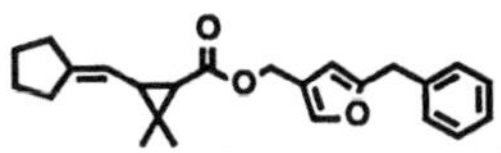	1 R trans	1	Bioethanomethrin	K'Othrin	Roussel-Uclaf
3	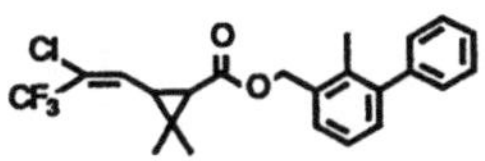	(±) cis, Z	2	Bifenthrin	Talstar	FMC
4	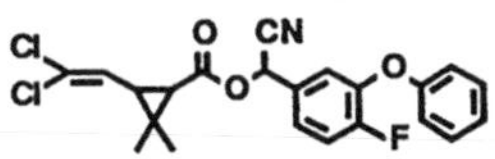	(±) cis, trans	8	Cyfluthrin	Baythroid	Bayer
		1 R cis α S 1 S cis α R 1 R trans α S 1 S trans α R	4	FCR 4545		Bayer
5	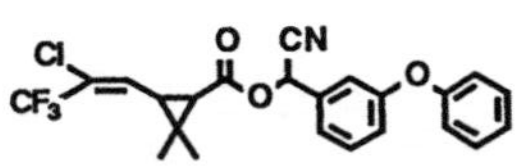	(±) cis, Z	4	Cyhalothrin	Grenade	ICI
6	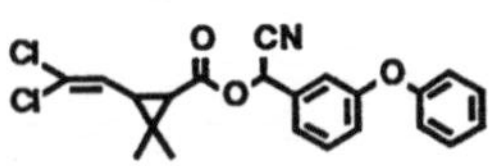	(±) cis/trans	8	Cypermethrin	Cymbush, Rip-cord, Ammo, Nural, Polythrin, Agrothrin	ICI, Shell, FMC, Dow, Ciba-Geigy, Sumitomo
		1 R cis α S				

7	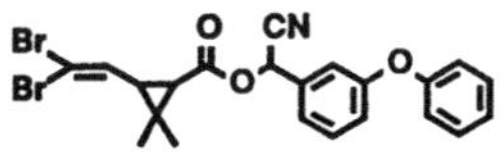	1 R cis α S	1	Decamethrin, Deltamethrin	Decis	Roussel-Uclaf
8	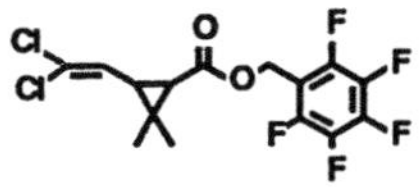	1 R trans	1	Fenfluthrin	Remedor	Bayer
9	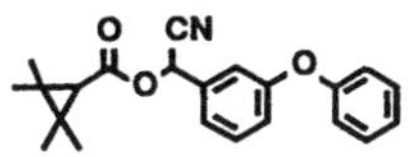	(±)	2	Fenpropanate, Fenpropathrin	Rody, Danitol	Sumitomo, Chevron, Shell
10	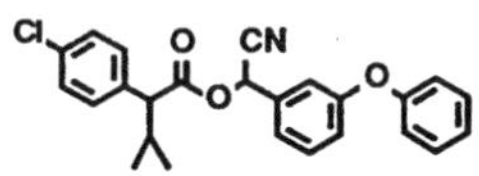	(±)	4	Fenvalerate	Sumicidin, Bell-mark, Pydrin	Sumitomo, Shell
11	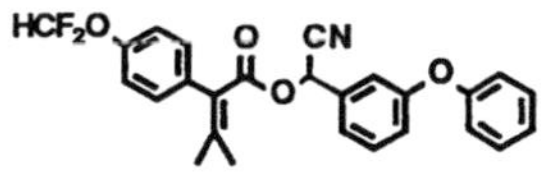	(+) S; α RS	2	Flucythrinate	Pay off, Cybold	ACC
12	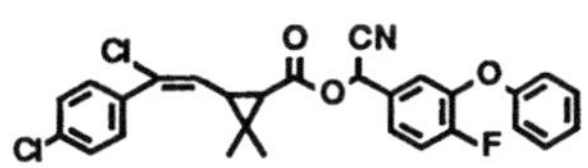	(±) cis/trans; Z, E	16	Flumethrin	Bayticol	Bayer
13	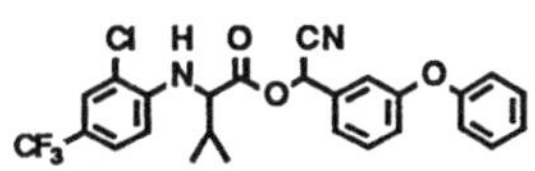	S; α RS	2	Fluvalinate	Mavrik	Zoecon
14	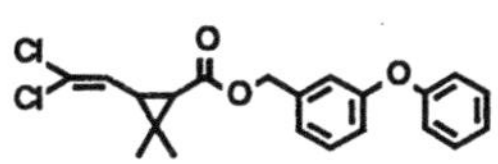	(±) cis/trans	4	Permethrin	Pounce, Am-bush, Talcord, Tornade	FMC, ICI, Shell, Rhone-Poulenc
15	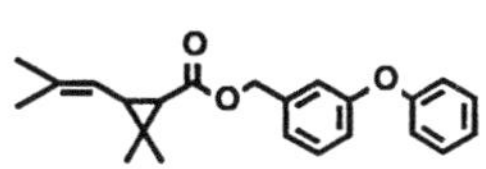	(±) cis/trans	4	Phenothrin	Walcide	Sumitomo

Table 1. (Continued)

Entry No.	Formula	Stereochemistry	No. of Isomers	Common name	Trade name	Basic producer, distributor
16	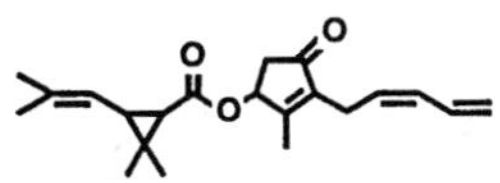	1 R trans; S	1	Pyrethrin I	Pyrethrum	"nature"
17	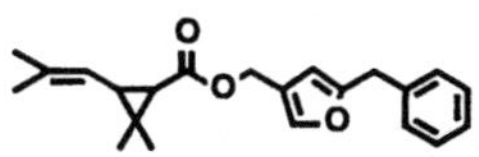	(±) cis/trans	4	Resmethrin	Chryson, Premgard	Sumitomo, FMC, a.o.
		1 R trans	1	Bioresmethrin	Resbuthrin	Wellcome, Fisons, a.o.
		1 R cis	1	Cismethrin		FMC, a.o.
18	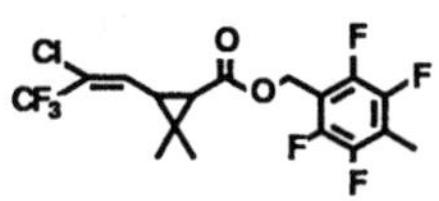	(±) cis Z	2	Tefluthrin	Force	ICI
19	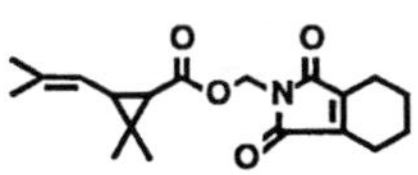	(±) cis/trans	4	Tetramethrin	Neopynamin, Phthalthrin	Sumitomo
20	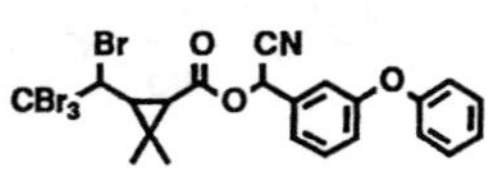	1 R cis, R´S´; α S	2	Tralomethrin	Scout	Roussel-Uclaf, Hoechst

vely labelled parent compounds and measuring and identifying the radioactivity remaining in the analytical samples after they have been exposed to field conditions ("weathering"). The problems — besides others — involved in determining bound residues should be mentioned here briefly: the methods of choice are in this case heating in hydrochloric acid solution under reflux or treatment with respective enzyme systems such as β-glucosidase.

Ref. [3]: Extract clean-up requires the greatest amount of time in the course of a residue analysis and the greatest skill as well. That is to say, the extract impurities, which trace analysts are so afraid of, may cause so-called "interferences" and must therefore be removed from the extracts as completely as possible. A number of most different procedures are used for this purpose — individually or combined. The main procedures are: Liquid/liquid partitioning (in case of pyrethroid analysis especially acetonitrile/n-hexane partitioning), steam distillation, "Sweep-co distillation" [6,7,8], molecular distillation [9] or different chromatographic procedures such as column chromatography (liquid/solid chromatography), preparative thin-layer chromatography, or gel permeation chromatography, as recommended by W. Specht [10] for residue analysis.

Ref. [4] and [5]: The final determination of the compounds in the extract is normally made by means of gas chromatography (GLC) or high performance liquid chromatography (HPLC). W. Blaß [11] discussed the question which of the two methods sould be preferred in a given case. Frequently, chromatographic columns of different polarity are used to substantiate positive findings. Even though the relative retention indices obtained in this way are by no means suitable to determine the "identity" of a substance — which is regrettably and incorrectly claimed by many authors — they still make a positive finding more likely. Retention times are physical characteristics just like a boiling point or a vapour pressure.

For the positive identification, spectroscopic methods must be used, possibly including mass spectroscopy (on- line coupled with GLC and/or HPLC). Only results "substantiated" with these methods will be justiciable.

For trace analysis of pyrethroids, the GLC method is definitely preferred. Only a few authors published HPLC methods so far. These will still be referred to later.

The reason for the dominance of the GLC method is the high sensitivity of the "electron capture detector" (ECD) to those pyrethroids containing halogen, which is fortunately applicable to most of these substances. The pyrethroids not containing halogen (chrysanthemate- and pyrethrate-type pyrethroids) must be detected with the far less sensitive and more universal "flame ionization detector" (FID).

Further general aspects of residue analysis of plant protectants have been covered by H. P. Thier and H. Frehse [12] as well as R. J. Hemingway [13], the latter mainly under the aspect of a cost-return-analysis of residue analytical work.

Going into details on individual analytical procedures would, of course, exceed the scope of this publication. Therefore, we will confine ourselves to indicating the respective PRINCIPLE OF THE METHOD. Details about the performance of the residue analyses may be taken from the respective publications cited in the bibliographgraphy, if required.

It should be mentioned that explanations and interpretations concerning the terms "limit of detection", "limit of determination", "limit of practical operating range" can be taken from Chapter 7 of the book written by H. P. Thier and H. Frehse [12].

2.1.1 Allethrin (Pallethrin), Bioallethrin

The chemical structure of the pyrethroids allethrin or pallethrin and bioallethrin is similar to that of the pyrethrins occuring in natural pyrethrum extract. The trace analytical determination of these substances causes particular problems, because heteroatoms such as chlorine, fluorine or nitrogen are lacking in the molecule and consequently the atom-selective gas chromatographic detectors like the halogen-sensitive ECD or the nitrogen-sensitive N-FID cannot be used.

Thus, we must depend on the more universal, but less sensitive FID or, in case of the use of HPLC, on the UV-detector which is also not highly sensitive nor highly selective. As a consequence of this the samples to be measured must be highly concentrated. This, in turn, may result in problems from interfering extract impurities which must be removed by means of time-consuming and expensive clean-up procedures.

In one case the quantification of allethrin residues using the ECD is described. For this purpose, W. Dicke and H. P. Thier [14] used SE-30-glass capillaries or DB-1-quartz capillaries as analytical columns. The investigations were made in the course of conducting a multiresidue analysis comprising the determination of 8 pyrethroids and the active ingredients of the pyrethrum extract in wheat, rye and their milling products. An acetonitrile-water mixture (66:33) was used for extraction of the active ingredients. The clean-up was made by partitioning against petroleum ether and gel permeation chromatography (GPC) using Bio-Beads S-X-3-gel and subsequent column chromatography with ®Florisil as adsorbent.

Incidentally, "multi-residue methods" are very much preferred by the residue analytical chemists working in official food quality control, because these methods allow the detection of entire groups of compounds such as the pyrethroids, necessitating only one single processing which facilitates the work considerably. This topic of "multi-residue methods" is, therefore, covered in a separate chapter of this publication.

Besides the above mentioned procedures a colorimetric method was described for bioallethrin. In that case J. M. Desmarchelier [15] suggested petroleum ether for the extraction of the pyrethroid from wheat, its milled products, polished rice, corn, sorghum, oats and barley. Preparative thin-layer chromatography was used for clean-up. After alkaline hydrolysis the resulting chrysanthemumic acid was derivatized with Deniges-reagent. Finally, the derivative was measured at 584 nm absorption band. Depending on the type of material to be investigated, the detection limits of the procedure were between 0.1 and 4 mg/kg which actually does not meet official requirements (see also under Sect. 3).

For the sake of completeness, two individual procedures should be mentioned which require great expenditures of appliances and/or special experiences of the analyst. These are GC-MS-coupling and immunological methods. The latter have also been described in the literature for allethrin and bioallethrin: K. D. Wing et al. [16], K. D. Wing and B. D. Hammock [17]. The possible stereo-selectivity of such immunochemical methods is an interesting aspect of this type of analysis. In addition to the already mentioned special experiences needed to perform this procedure, two other disadvantages should be mentioned in connection with the immunochemical analysis of pyrethroid residues:

1. The need to work in aqueous solutions may cause problems with the highly lipophilic pyrethroids and

2. the development of an immunological test takes a rather long time (approx. 1/2 year) which, in turn, causes costs that can only be justified if subsequently a large number of samples have to be analysed.

For the listed reasons it seems rather unlikely that immunological methods of trace analysis also of pyrethroids will find more acceptance on a large scale than the conventional methods. It is conceivable, however, to use an immunoassay as a kind of pyrethroid-screening-test which allows a yes-no-decision. If such a test would result in the decision "yes", it would at least give a qualitative hint to the presence of a pyrethroid residue which would then have to be analysed using a conventional method.

2.1.2 Bioethanomethrin

This pyrethroid is only rarely mentioned in the literature. It is a halogen- and nitrogen-free substance so that the general information given under allethrin and bioallethrin also applies in this case.

D. A. George et al. [18] described derivative-GLC for the pyrethroid after saponification to the respective alcohol and to the acid compound. The alcohol was converted with trichloracetyl chloride to the corresponding ester while the acid was esterified with trichloroethanol. The halogen content in the derivatives allows the determination by means of GLC-ECD. The analytical result was substantiated by the synthesis of the two derivatives.

The method was tested with lentils. Dichloromethane was used for extraction of the active ingredient. The clean-up of the extract was achieved by column liquid chromatography using alumina as an adsorbent and that of the lentils after liquid-liquid-partitioning between acetonitrile and n-hexane.

2.1.3 Bifenthrin

Also on this synthetic pyrethroid the information found in the literature is scanty [19], anonymous), the reason being that the development of this acaricide is not yet completed. Even though an analytical method was not yet published, actually only the GLC-ECD determination is to be considered for bifenthrin due to its fluorine and chlorine content in the molecule, as is the case with the other halogen containing pyrethroids.

2.1.4 Cyfluthrin

The first reports about methods to analyse for residues of this synthetic pyrethroid appeared in the early eighties. These concerned the determination of cyfluthrin in animal tissues and milk as well as in cereals, tea and soil: K. Wagner [20,21]; R. R. Gronberg and L. K. Pfankuche [22]; H. R. Shaw et al. [23]. Gronberg/Pfankuche and Shaw et al. included the main metabolites into the analytical determination in addition to the parent compound. The latter is not the case with the work done by H. R. Shaw [24] who included eggs into his investigations, as well as by A. M. Harbin [25] and Y. Ishii/J. Ueyama [26] who analysed exclusively for parent compound in samples originating from plants and soil. The publication by Ishii/Ueyama [26] is interesting because these authors used a nitrogen-selective detector (N-FID) for the final determination. In all other studies on cyfluthrin so far mentioned it was always the halogen-sensitive electron capture detector which was used.

K. Riegner [27] determined cyfluthrin residues in sediment and A. Inkmann-Koch [28] described a method for the detection of the substance in fish. R. R. Gronberg et al. [29] determined cyfluthrin residues gas chromatographically in various foodstuffs-also after processing: bread, butter, flour, "hambruger" (raw), head lettuce, macaronies, milk, peaches, rice and sugar were included into the investigations. W. Blaß [30] presented a similarly comprehensive method description for the analysis of residues of unchanged cyfluthrin in numerous plant materials, which are listed in detail in part II on residue data.

In all studies mentioned the parent compound or the metabolites were extracted with organic solvents or mixtures of organic solvents. The preliminary clean-up is often followed by a liquid-liquid-partitioning between acetonitrile and n-hexane. The extract clean-up is then made by means of column chromatography using Florisil or standardized silica gels or also using alumina. Quite frequently the extracts are evenly subdivided and the aliquots are cleaned up using mini silica gel columns or disposable cartridges. Gel permeation chromatography (GPC) is also used for extract clean-up.

For the determination of the residues in the cleaned-up extracts the electron capture detector (ECD) is used with only the one a.m. exception [26]. In most cases packed columns are recommended for separation, in which the support materials are coated with methyl or mixed alkylated polysiloxanes as stationary phases. Examples for such polysiloxanes are: DC-200, SE-30, OV-1, OV-101, SE-52, SE-54, OV-17, OV-1701.

Capillary column GLC may also be considered [27]. The capillary column GLC has the advantage of greater separating efficiency and allows the gas chromatographic separation to be performed at largely inert surfaces especially if quartz glass capillaries are used — yet it has the disadvantage of a lower capacity compared to the packed columns.

All methods must, of course, be checked in fortification tests by applying known concentrations of the parent compound to untreated control material. However, such experiments only give a rough idea of the quality of an analytical method, as they would not make sure that "grown residues" are also extracted. The problems involved in

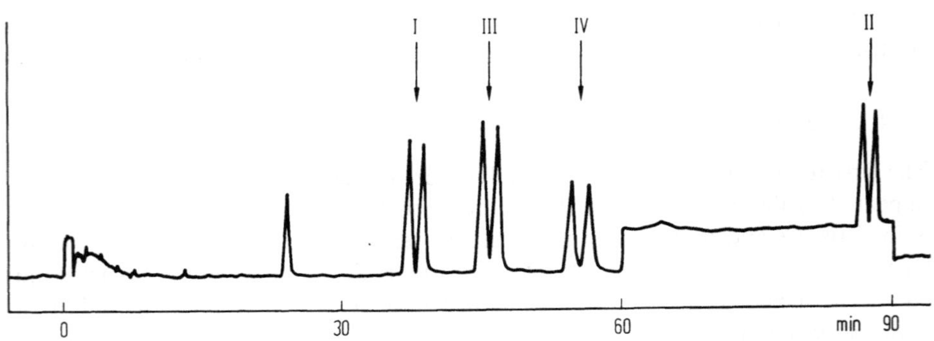

Fig. 1. Separation of the eight optical isomers of CYFLUTHRIN using HPLC column switching and the BAKERBOND (J. T. Baker Co.) chiral columns (R)-N-3.5-dinitrobenzoylphenylglycin and (S)-N-3.5-dinitrobenzoylleucin both covalently bonded to modified 5 μm silica (according to J. Krohn (32))

I: 1R-3R-αR + 1S-3S-αS } 1.3 cis
II: 1R-3R-αS + 1S-3S-αR }
III: 1R-3S-αR + 1S-3R-αS } 1.3 trans
IV: 1R-3S-αS + 1S-3R-αR }

analysing bound residues were mentioned already previously. In this case only the use of radioactively labelled compounds and the analytical determination of the resulting grown residues lead to reliable evidence.

On the other hand it must be taken into consideration that bound residues are not necessarily bioavailable. And, of course, non-bioavailable residues do not constitute a risk to the consumer. So far, the knowledge about the aspect of the bioavailability of pyrethroids is only scanty. The topic "bioavailability of bound residues" was covered a.o. in a publication by S. U. Khan [31].

A successful separation of cyfluthrin enantiomers should still be mentioned briefly. The determination of the ratio of enantiomers is important in the interest of a uniform quality of the active ingredient. But also the residue analysis of defined enantiomers could well be of interest. J. Krohn [32] described the separation by means of two columns of the same chiral stationary phase (see Fig. 1). A disadvantage is the relatively long time (95 min.) needed for the analysis. Incidentally, for separation of the four diastereomers of cyfluthrin Krohn uses a precolumn which is coupled on-line with the subsequent chiral columns via a column switching system. Cyfluthrin was also incorporated into the "Multi-residue methods" (see at the end of this chapter): W. Dicke and H. P. Thier [14] and R. A. Chapman [33].

2.1.5 Cyhalothrin

A. Jacques [34] described a residue method for animal tissues and milk. This method is based on GLC-ECD detection using a pulsed 63Ni-detector, which is common practice nowadays. Prior to the introduction of the easier to handle 63Ni-ECD's, detectors were used which were based on tritium as source of β-rays, in which case this radiation source producing high velocity electrons actually consisted of titanium tetratritide. Also in this method, column chromatography using Florisil as adsorbent is recommended for extract clean-up. A conventional packed column with OV-101 as stationary phase is recommended for use as gas chromatographic column. The limits of detection of the method — preferably cefined as "limits of practical operating range" — are 0.01 mg/kg for animal tissues (liver, muscular substance, kidney, fat of various origin) as well as 0.002 mg/kg for milk.

The use of cyhalothrin only in cattle dips explains the fact that A. Jacques limited his efforts to animal tissues and milk when developing the analytical method.

2.1.6 Cypermethrin, Alphamethrin

The two compounds differ from each other only with regard to the isomer composition (see Table 1). Therefore, the discussion of the analytical methods for both substances is summarized in one chapter.

Besides deltamethrin, fenvalerate and permethrin, cypermethrin belongs to the most thoroughly investigated pyrethroids as far as residue analytics are concerned.

Numerous publications originate especially from the Far East, where approx. 30% of the pyrethroids used in the world are applied.

In accordance with the wide use of cypermethrin, a great number of analytical methods have been published. These cover all kinds of plant and animal materials as well as soil and water. Thus, R. A. Chapman [33] and R. A. Chapman and C. R.

Harris [35] reported about isomer separation in residue of cypermethrin and other pyrethroids by means of HPLC via chiral stationary phases and/or by means of gas chromatography after formation of the 1-menthylester of the respective cyclopropane carbonic acid and GLC using OV-210 as stationary phase. Epimerization in C-1 or C-3 position was not observed. The limit of detection achieved in soil was reported to be 0.05 mg/kg. This type of trace analysis gives an indication of the speed of degradation of certain enantiomers which can be quite different.

A method for the determination of cypermethrin in celery and animal products (muscular substance, yolk and milk) was published by H. E. Braun and J. Stanek [36].

A. Sapiets et al. [37] and S. Forbes and A. J. Dutton [38] determined cypermethrin residues in various fruits and vegetables as well as milk, soil and water. L. P. Sarna and G. R. B. Webster [39] described a method for the determination of residues in stored wheat and its milled products.

Common to all the methods mentioned in the last paragraph is the extraction with acetone/hexane or acetonitrile. In one case [36] the extraction with acetonitrile is followed by a clean-up using liquid-liquid-partitioning between acetonitrile and n-hexane.

A. J. Dutton [38] "miniaturized" his method. He used disposable cartridges (®Bond-Elut/CN-Phase, 50 mg sorbent amount for extract clean-up). This measure saves mobile solvent. The use of smaller amounts of solvents, in turn, results in a saving of time not only with the elution but also when extracts are to be concentrated to smaller volumes. Moreover, Dutton combined the extraction and the subsequent liquid-liquid-partitioning between water/acetonitrile/hexane in one analytical step, a measure which also helps to save time and money.

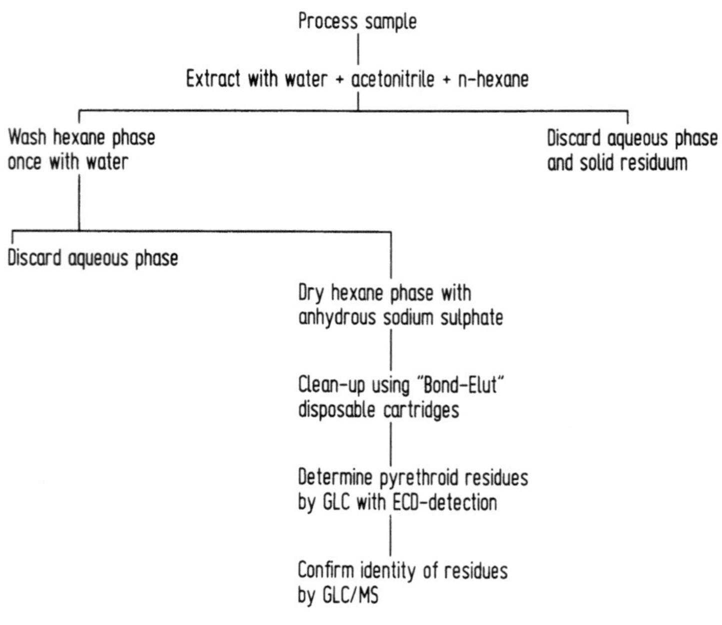

Fig. 2. Scheme for the determination of CYPERMETHRIN and ALPHAMETHRIN residues in crops and soils (according to S. Forbes et al (38))

Miniaturization techniques of this kind should be standard procedure in every modern trace analytical laboratory.

Braun and Stanek [36] pointed out that the cited processing techniques may be used simultaneously for the determination of insecticidal chlorinated hydrocarbons. It remains to be mentioned that with all cited methods packed columns with the already mentioned stationary phases and the pulsed 63Ni-ECD are used for detection.

A. Sapiets [37] indicated additionally that a separation of enantiomer pairs may be achieved by means of quartz capillaries (25 m OV-101/WCOT).

The limit of detection reported for all the methods cited were in the range of less than 0.005 to 0.02 mg/kg material.

L. Kadenczki et al. [40] presented a "new extraction procedure": after preliminary blending of the analytical sample it was ground with Florisil in a mortar. The resulting powder was then transferred into a chromatographic column. The active ingredient was extracted from this column with ethyl acetate or a dichloromethane/acetone mixture (90:10), respectively.

Residues of alphamethrin in cereals and soil may be analysed by means of the method of S. Forbes et al. [38] which was already mentioned.

As this miniaturized method may offer the possibility to determine also other pyrethroids, the scheme of analysis is presented in Fig. 2.

2.1.7 Deltamethrin (Decis)

An earlier name of this compound is decamethrin. Still today deltamethrin can be found in the literature under this older common name.

Since early 1982 several analytical methods have been published in technical literature, the principle of which is briefly outlined in the following: M. H. Akhtar [41] described a gas chromatographic method which allows to analyse biological samples — urine, feces, milk, liver, kidney, muscular substance and subcutanoeous fat — for possible residues of deltamethrin. The extraction was made with acetone with two exceptions: milk and urine were extracted with hexane. The preliminary clean-up was performed by liquid-liquid-partitioning against acetonitrile while the actual extract clean-up was made by means of a microcolumn. This column — a "Pasteur pipette" — was packed to a level of 5 cm with Florisil which had been deactivated with 10% of water before-hand. The Florisil phase was layered with 3 cm of a mixture of 25% cellulose in activated charcoal.

The gas chromatographic measurement was performed with a 63Ni-ECD after chromatography using SE-30 in a packed column. The recovery rates of the method, mesured after addition of active ingredient to the untreated material in concentrations between 0.01 and 0.1 mg/kg, were between 67 and 104% of the amounts added, as is common with this type of trace analytical methods. The "limit of practical operating range" was 0.01 mg/kg.

In connection with a multi-residue method, R. Agneessens et al. [42] described an interesting extraction and extract clean-up procedure for deltamethrin, permethrin and 17 other insecticides and fungicides in plant materials. As all active ingredients contained halogen, the final determination was performed with the 63Ni-version of the electron capture detector (ECD). The chromatography was performed using a thin-film capillary (0.15 μm film thickness) of a length of 25 m and having an i.d. of 0.32 mm, coated with OV-1 as the stationary phase.

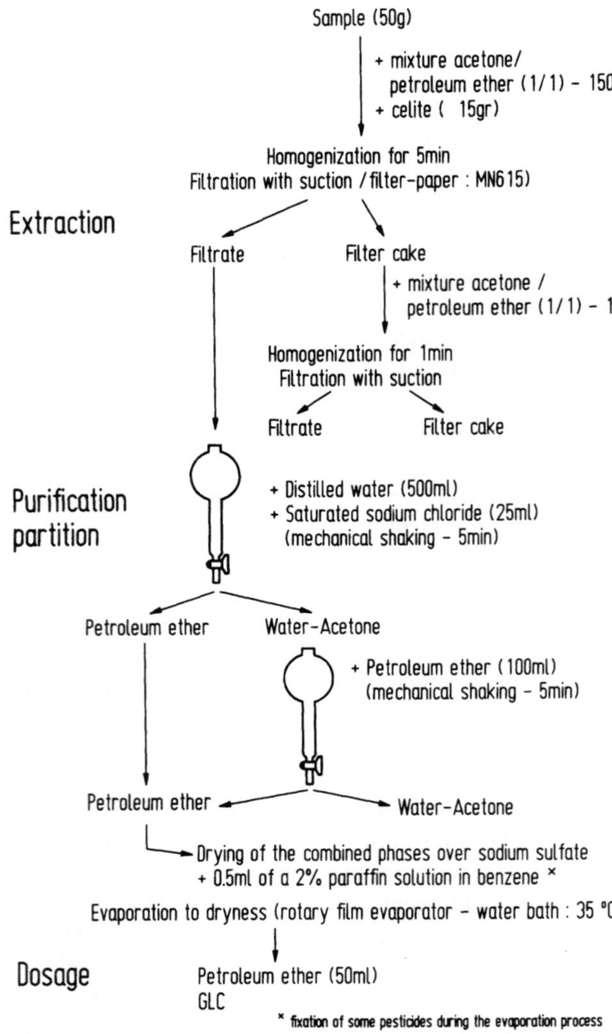

Fig. 3. Extraction and purification of pesticide residues from crop extract: schematic description (according to R. Agneessens et al. (42))

Extraction and clean-up according to Agneessens et al. [42] are presented schematically in Fig. 3.

A very comprehensive report on methods was published by M. Vaysse et al. [43]. Fatty and non-fatty plant materials, soil, milk and animal tissues were included in the method. The extraction was made with unpolar organic solvents such as petroleum ether and hexane, and in case of soil and animal tissues with mixtures of acetone/hexane or petroleum ether/ethylether respectively.

Extract clean-up procedures to be considered were: liquid-liquid-partitioning, column chromatography using Florisil or gel permeation chromatography (GPC).

The final analysis was made gaschromatographically with a 63Ni-ECD using conventionally packed columns (1% SE-30).

The studies of C. Meinard et al. [44, 45] could be quite interesting for the residue analyst. The authors worked on the determination of the diastereoisomers of deltamethrin with HPLC/UV-detection while the enantiomers may be identified with HPLC- polarimeter-coupling. These studies are of importance considering the partial isomerization of deltamethrin under the influence of light.

By coupling HPLC with a liquid scintillation detector, C. Meinard [45] answered the question whether the insecticidal activity of tralomethrin is based on its transformation into deltamethrin. The answer was: "no". The method allows the differentiation between the diastereo-isomers via their retention times and the quantitative determination on the basis of their radioactivity.

2.1.8 Fenfluthrin

A residue analytical method for the determination of this substance in foodstuffs originating from animals (liver, kidney, fat, muscular substance and milk) was presented by W. Blaß and K. Riegner [51]. Following this method, the active ingredient was extracted from the analytical material with acetonitrile (except for milk). This was followed by liquid-liquid-partitioning between acetonitrile and hexane for preliminary clean-up, which removed the lipids from the raw extract. The extract was then actually cleaned up by column chromatography using silica gel.

The subsequent quantitative determination of the active incredient was performed by gas chromatography using an electron capture detector. Conventionally packed columns with SE-30 as stationary phase served as separation system.

The limits of detection of this method are between 0.01 and 0.02 mg/kg, for milk at 0.005 mg/kg.

Milk was extracted with acetone. Prior to the preliminary clean-up by means of acetonitrile-hexane-partitioning, the residues were transferred into dichloromethane in order to remove the polar extract impurities with the acetone-water phase.

2.1.9 Fenpropanate (Fenpropathrin)

This pyrethroid is of particular interest as acaricide but is also effective as soil insecticide. It should be noted that its acute oral toxicity (LD50 for male and female rats of 164 and 107 mg/kg resp.) is relatively high compared to the other pyrethroids.

Y. Takimoto et al. [46] described an analytical method covering many kinds of fruits and vegetables as well as green tea. Following this method, analytical materials with a high content of water were extracted with acetone, i.e. with a polar solvent. Materials with a low water content such as green tea were to be extracted with a methanol-water mixture. After transfer of these raw extracts into unpolar solvents (dichloromethane or n-hexane, resp.) the extracts were cleaned up with silica gel or Florisil. The final determination was performed by means of a 63Ni-ECD after chromatography using a packed SE-30 column. Limits of detection were indicated to be between 0.001 and 0.02 mg/kg. Interferences from other, frequently used pyrethroids (such as permethrin, cypermethrin, deltamethrin, phenothrin and fenvalerate) were not observed, as fenpropanate had a shorter retention time under the gas chromatographic conditions employed than the other five pyrethroids mentioned. Fenpropanate is also covered in "Multi-residue methods" [14, 33, 47, 49, 50]. These methods will be discussed more in detail at the end of this chapter.

2.1.10 Fenvalerate

B. D. Hill et al. [52] determined fenvalerate residues in alfalfa. After extraction of the analytical material with an acetone-hexane mixture (1:1), water was added. The resulting hexane phase was processed further. The extract was cleaned up by means of a "tandem-micro-column-system" with alumina and silica gel, both deactivated with 6% of water, being used as chromatographic adsorbents. The gas chromatographic determination was performed with a pulsed 63Ni-ECD. The GLC-column was a conventionally packed column with OV-210 as the stationary phase.

In the course of a ring trial to determine residues of fenvalerate, five different laboratories investigated processed products of apples with the respective byproducts (applesauce, apple juice, apple pomace, apple peels and apple cores). The results were evaluated by T. D. Spittler [53].

The gas chromatographic determination of the residues by means of a 63Ni-ECD using packed columns (stationary phases: OV-101, OV-17, SP-2100 or ®Dexsil) was preceded by column chromatographic extract clean-up using Florisil-®SepPak-cartridges.

T. D. Spittler [54] also determined fenvalerate residues in processed products of tomatoes and respective byproducts. Hexane-isopropanol (3:1) served as a solvent system for extraction of the active ingredient. Extract clean-up and final determination were performed as briefly described under [53].

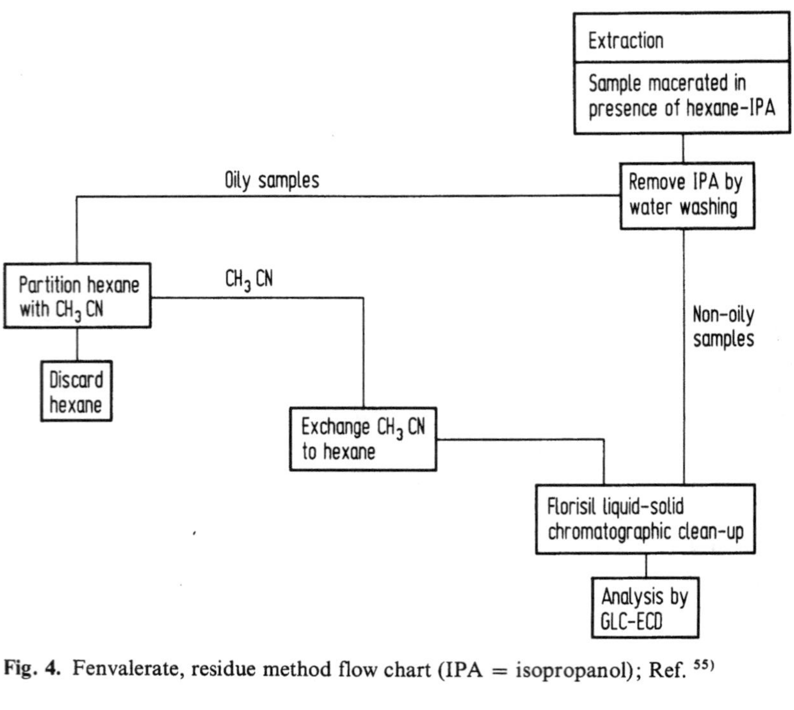

Fig. 4. Fenvalerate, residue method flow chart (IPA = isopropanol); Ref. [55]

Details concerning the residue analytical determination of fenvalerate in samples originating from plants, water, soil and animals can be taken from a report of Shell Co [55]. The principle of the method is presented schematically in Fig. 4.

Residues of fenvalerate in milk can be determined using the method of E. Papadopoulou-Mourkidou [56]. This method allows the separation of the optical isomers of fenvalerate by means of HPLC on a chiral column. This column contains (R)-N-3,5-dinitrobenzoyl-phenyl-glycine (DNBPG), covalently bound to aminopropyl silica gel (particle size 5 μm). The phase is sold by Baker Corp. as ®Bakerbond. The isomers were eluted with mixtures of methanol/isopropanol/n-hexane.

In the course of the analytical procedure first the milk proteins were precipitated with acetonitrile and filtered off. The fenvalerate isomer mixture was extracted from the filtrate with n-hexane. As only one of the 4 isomers (the one with SS-configuration)

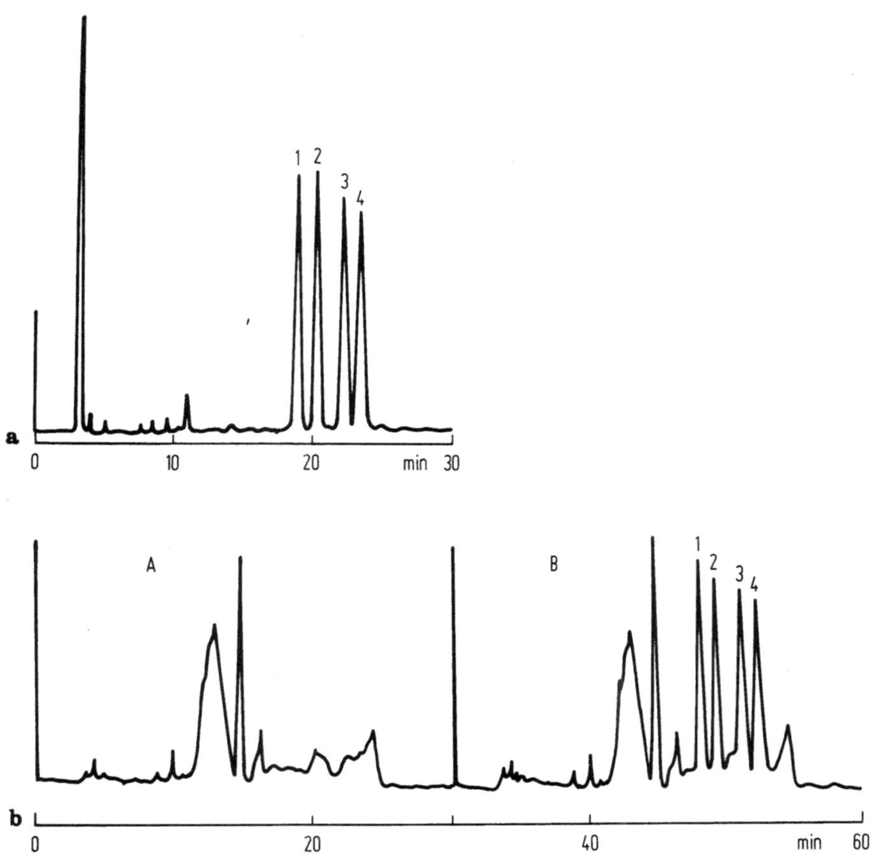

Fig. 5a. Sample chromatogram from analysis of Pydrin®. Dilution, 0.33% (v/v) in hexane. Column, BAKERBOND chiral column. Eluent, 2-propanol:hexane (0.1:99.9, v/v) flow rate, 1 ml min^{-1}. Detection at 254 nm and 0.1 AUFS. Peaks identified as 1, 2, 3, and 4 correspond to four optical isomers of fenvalerate. **b.** Sample chromatograms from analysis of control (A) and fortified (B) milk samples. Fortification level 5 ppm with fenvalerate. Column, BAKERBOND chiral column. Eluent, methanol:2-propanol: hexane (0.1:0.3:99.6, v/v/v) flow rate, 1 ml min^{-1}. Detection at 280 nm and 0.02 AUFS. Peaks identified as 1, 2, 3, and 4 correspond to four optical isomers of fenvalerate.

is effective as active ingredient, it is interesting to know the optical purity of fenvalerate formulations, which can also be determined analytically with this method. Figures 5a and 5b show the quality of the separation of the four optical isomers of fenvalerate in milk.

Further hints concerning the methodology of fenvalerate residue analysis can be taken from the part "Multi-residue methods" at the end of Chap. 2.1.

2.1.11 Flucythrinate

Flucythrinate is a broad spectrum insecticide. C. Cordon [57] described the determination of residues of this substance in various plant materials (cabbage, broccoli, apples, wheat, barley and hops). The extraction solvent was dichloromethane, in case of hops it was petrolether. Florisil column chromatography was used for extract clean-up.

An interesting detail of this method is that the author recommended the removal of interfering extract impurities by "freezing out" (—6 °C).

The final determination of the residues by gas chromatography was performed with a conventionally packed column (stationary phase: OV-1) and a 63Ni-ECD.

Two signals (peaks) resulted due to the presence of 2 diastereomer pairs of which the substance is composed (RR/SS/RS and SR).

The limit of practical operating range was indicated to be 0.05 mg/kg for all analytical materials mentioned.

2.1.12 Flumethrin

This substance is used in the veterinary field for the control of ticks. Moreover, it is useful to control mites on honeybees.

Without giving analytical details, H. Neuhauser [58] reported in a synoptic article about the residue situation in foodstuffs deriving from animals after application of this compound. Muscular substance, fat, liver, kidney and milk of dairy cows were investigated. The limit of detection of the method was indicated to be 0.05 mg/kg and 0.01 mg/kg — the latter in case of milk, based on milk fat.

2.1.13 Fluvalinate

R. B. Leidy et al. [59] determined fluvalinate in tobacco besides permethrin. They used a HPLC-method (column packing: ®RadicalPak-C18 rp), mobile solvent system: acetonitrile-water-isopropanol (2:1:1).

A comprehensive method report covering analytical materials originating from plants and animals as well as water and soil was published by W. L. Fitch et al. [60]. The method is based on gas chromatography with ECD and quartz capillary columns. The authors pointed out that the substance, which is relatively polar compared to other pyrethroids, is unstable on packed gas chromatographic columns and that therefore such columns are not suitable for chromatography of this substance.

Depending upon fat and water content of the material to be analysed, different solvent mixtures were used for the extraction of the active ingredient.

For extract clean-up, liquid-liquid-partitioning, Florisil column chromatography or gel permeation chromatography (GPC) were used. Careful extract clean-up is essential especially if "narrow-bore" capillaries are used because these capillaries have a very limited sample capacity. These correlations were also mentioned by Fitch in his publication. If such capillaries were used, the author recommended the "internal standard method" for the quantitative evaluation of the chromatograms in order to correct for losses during extraction or extract clean-up, but also to compensate the variability with capillary column injections due to substance discrimination or changes of the split ratio.

For GC-determination of fluvalinate, the 3-phenoxybenzyl-2-(4-trifluoromethyl)-anilino)-3-methylbutanoate, which has a structure very similar to that of fluvalinate, has proven to be well suitable as an internal standard compound. A known amount of this was added to the analytical sample at the start of the processing, as is common practice with the "internal-standard-method".

2.1.14 Permethrin

Permethrin belongs to the pyrethroids being most widely used. Consequently there are many publications on residue analytical methods and the residue situation with this compound.

Thus, A. Mac Phee et al. [61] detected residues of permethrin in blueberries after application of 0.4 kg a.i./ha. The active ingredient was extracted with acetone. After transfer into hexane for removal of polar extract impurities with the acetone/water phase, the extract was cleaned up by means of column chromatography using Florisil. The gas chromatographic determination was performed with an ECD and a packed column (OV-210 as stationary phase). The limit of detection of this procedure was indicated to be 0.01 mg/kg.

E. Papadopoulou-Mourkidou et al. [62] described an online method for permethrin using HPLC/IR-detection. This was the first reported coupling of equipment of this type in the field of pesticide residue analysis. Acetone was used to extract the active ingredient from head lettuce. This was followed by acetonitrile-hexane-partitioning for preliminary extract clean-up. After filtration through SepPak-cartridges the separation was performed by HPLC. The packing of the column for this was ®Partisil with a particle size of 5 µm. The cis/trans-isomer separation was achieved with the mobile phase system 1-tetradecene-dichloromethane-cyclohexane (3:70:27). The IR-determination was made at 5.8 µm and 0.025 AUFS (= absorbance units full scale).

The limit of detection of this procedure was indicated to be 0.2 mg/kg.

The extracts had to be highly concentrated by evaporation: for measurement, 100 g fresh head lettuce were reduced to a final volume of 0.2 ml. The high concentrating was indispensable due to the relatively low sensitivity of the IR-detector.

In connection with a multi-residue method which allowed to determine traces of approximately 100 active ingredients, mainly belonging to the groups of chlorinated plant protectants and organophosphates, H.-J. Stan and D. Mrowetz [63] determined also permethrin analytically. They used 2-dimensional gas chromatography with capillary columns and parallel detection by means of a flame photometric detector (for the organic phosphates) and ECD (for the chlorinated plant protectants). The first column contained methyl polysiloxane as stationary phase and was used as a kind of "screening"-column while the second column, the main column for separation, contained phenyl-cyanopropyl-modified methyl polysiloxane as the stationary phase.

In Fig. 6 the respective column switching ("live switching" of Siemens) is presented. Column switching techniques of this type can be used in trace analysis to a great advantage. These techniques which are incidentally also applicable in connection

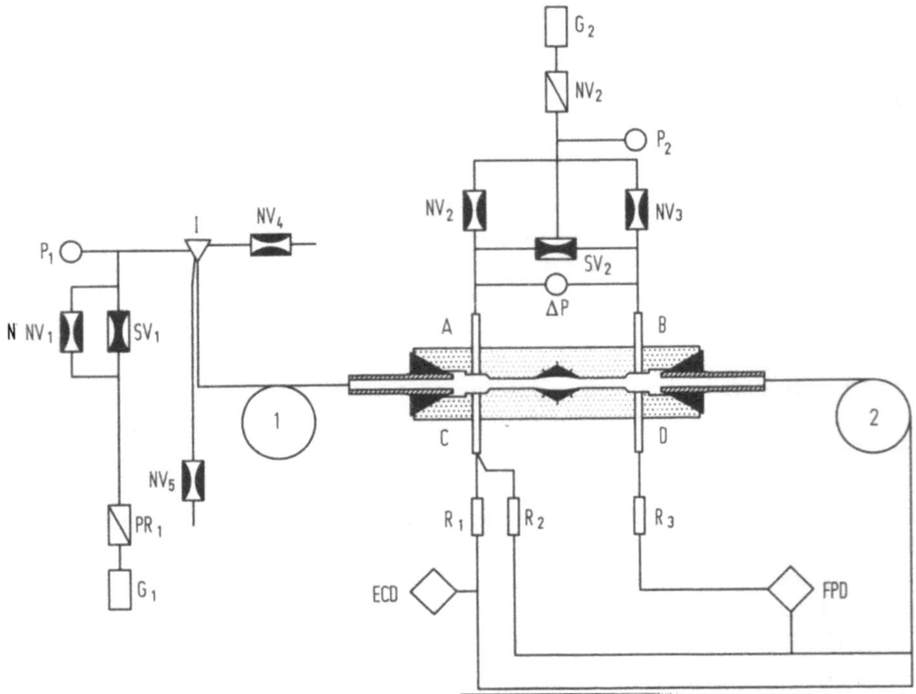

Fig. 6. Flow diagram of the gas chromatographic system. G = gas supply; PR = pressure regulator; P = pressure gauge; SV = solenoid valve; NV = needle valve; ΔP = differential pressure gauge; R = restriction; I = injector; FPD, ECD = detectors; A, B = switch gas entry; C, D = effluent exit.

with HPLC, provide important additional information within a relatively short time of analysis: columns of varying polarity yield different retention indices which confirm an analytical result. The same applies to the parallel operation of different atom-selective detectors. These correlations were discussed by W. Blaß et al. [64]

Capillary-GLC was also recommended by N. V. Fehringer and S. M. Walters [65] for "tentative identification" of pesticides. Besides many other active ingredients they also described the detection of cis/trans permethrin.

The authors discussed in detail the technical requirements to be met when working with the capillary column technique. They pointed especially to the relatively low capacity of such columns (extract impurities!) which necessitate time-consuming extract clean-up techniques.

A HPLC-method for the determination of permethrin and fluvalinate in tobacco was published by R. B. Leidy et al. [59]. (Details may be taken from the Chap. 2.1.13).

The separation of the 4 isomers of permethrin using a HPLC-method was described by K. M. S. Sundaram et al [66]. The column packing consisted of "ODS" (= octadecylsilyl), 5 µm particle size, bound phase. Mobile solvent: acetonitrile-water (70:30, v/v), UV-detection at 200 nm. This method was applied to investigate impact separator and water samples. The limit of detection was 1–3 ng permethrin, absolute.

H. E. Braun and J. Stanek [36] determined permethrin, cypermethrin and fenvalerate in celery, yolk, beef and milk. This is a GLC-method with ECD-detection. The extract clean-up was performed with Florisil column chromatography. The limit of detection was 0.005 mg/kg. Further details can be taken from Chapter I, 6 on cypermethrin. The authors pointed out that with this method chlorinated hydrocarbon insecticides can be determined simultaneously. Incidentally, the quantitative determination was made in this case preferably with OV-1 as a stationary phase in packed columns. A separation of isomers was not possible with this method. It yielded only 1 signal (peak), which facilitated the quantitative determination.

Further hints to analytical methods for the analysis of traces of permethrin can be taken from Chap. 2.1.

2.1.15 Phenothrin

Phenothrin — the spelling fenothrin is also being used — is a pyrethroid without heteroatoms in its molecule and is therefore difficult to detect and identify trace analytically in non-derivatized form. Therefore, Y. Takimoto et al. [67] described the derivatization of the substance after alkaline hydrolysis of the ester bond and esterification of the resulting 3-phenoxybenzyl alcohol with 2,4-dichlorobenzyl chloride. The derivative was determined gas chromatographically with an ECD.

The active ingredient was extracted with water-methanol-acetonitrile (2:1:4 v/v/v) in case of materials of high water content and with methanol in case of material of low water content. Preparative thin-layer chromatography and column chromatography with Florisil or silica gel resp. were recommended for extract clean-up. The final gas chromatographic determination was performed with packed columns (stationary phases: OV-101 and ®Apiezon L).

The method was tested in fortification experiments on cabbage, green peppers as well as rice grains and straw. In these experiments, recovery rates of 80–88 % of the amounts of active ingredient resulted with the added amounts having been between 0.1 and 1.0 mg/kg.

The limit of detection of the method was indicated to be 0.01 mg/kg for rice straw and 0.005 mg/kg for the other materials. The authors pointed out that gas chromatography using the flame ionization detector (FID) may be applied in case of higher residue levels.

2.1.16 Pyrethrins

Even though the pyrethrins (pyrethrin, cinerin, jasmolin, I and II each) are also components without heteroatoms, various gas chromatographic methods of determination have yet been published which are based on ECD-detection.

Thus, H. Siltanen et al. [68] reported on residues of pyrethrins in soil, measured with an electron capture detector using packed columns or capillaries. For the extraction of the active ingredient the authors used Soxhlet extraction with a hexane-acetone mixture (90:10). The limit of determination of the procedure was 0.1 mg/kg.

Also in the course of applying the multi-residue methods of W. Specht [10] and of W. D. Weinmann, H.-G. Nolting and J. Siebers [69], the pyrethrins were mesured with GLC-ECD. Further details concerning these methods can be taken from Chap. 2.1.

Incidentally, pyrethrins are normally used together with the synergist piperonyl butoxide. Residues of this substance must also be analytically determined, as Maximum Residue Limits were set for it in various countries. For this purpose, W. Specht [10] applied GC-MS-coupling, while the authors cited under [69] determined the piperonyl butoxide using gas chromatography with ECD after bromination with a bromine-iodine-mixture.

T. Ando et al. [70] described the separation of stereo-isomers by means of HPLC using Nucleosil 5 CN. The mobile solvent in this case was 1% tetrahydrofurane in n-hexane.

2.1.17 Resmethrin, Bioresmethrin, Cismethrin

These substances differ from each other only by the composition of isomers. In recent years no additional publications appeared concerning the analysis of these compounds so that the status in this respect as reported by E. Papadopoulou-Mourkidou [4] is still the actual one. As these substances do not have any heteroatoms in their molecules, the atom-selective gas chromatographic detectors are not applicable for the final determination of the unchanged parent compounds. J. M. Desmarchelier [71], therefore, recommended alkaline hydrolysis followed by colorimetric determination of a derivative of the chrysanthemumic acid which resulted from that hydrolysis. The derivatization reagent was acidified Hg-II-sulfate.

The procedure was tested with wheat, barley, oats, sorghum and rice, the latter also after processing.

According to J. M. Desmarchelier, the method [15] mentioned under 1. Allethrin/Bioallethrin is applicable to residues of bioresmethrin as well. Some further hints concerning analysis of traces of these substances may be taken from Chap. 2.1.

2.1.18 Tefluthrin

In a publication on the uptake and degradation of tefluthrin in corn and sugar beets, R. Amos and J. P. Leahey [72] also reported about some details of the trace analytical determination of this substance. The active ingredient was extracted with acetonitrile. This was followed by extract clean-up by means of liquid-liquid-partitioning using hexane, dichloromethane or ethyl acetate. The chromatographic determination was performed by thin-layer chromatography. Capillary GLC/MS-coupling was used to confirm the results.

In this publication the problems associated with "bound residues" were discussed in detail.

2.1.19 Tetramethrin

According to J. M. Desmarchelier the method [15] cited under 1. Allethrin/Bioallethrin can also be applied to analyse for tetramethrin residues.

2.1.20 Tralomethrin

C. Meinhard et al. [45] published an analytical method which is based on coupling of HPLC and a liquid scintillation counter (^{14}C). The principle of the method was reported already in Chap. 2.1.7.

Multi-residue Methods

These methods, also called multi-methods or multiple analytical methods, are of great interest for official laboratories being involved in food quality control analysis for consumer protection. Quite frequently, the "needle in the haystack" must be searched

Table 2. Multi-residue methods for pyrethroid insecticides

Compounds	Materials to be analysed	Principle of the final measurement	Ref.
Allethrin, Cyfluthrin, Cypermethrin, Deltamethrin, Fenpropathrin, Fenvalerate, Permethrin, Tetramethrin and Pyrethrum	wheat, rye and milled products	GLC-ECD	14)
Cypermethrin, Alphamethrin	crops and soils	GLC-ECD	38)
Bioethanomethrin, Permethrin	lentil foliage, lentils, green peas, and cabbage	GLC-ECD after derivatization	18)
α-Cyano-3-phynoxybenzyl-pyrethroids: Fenvalerate, Fenpropanate, Cypermethrin, and Decamethrin	confirming the identity	GLC-ECD after derivatization	47)
Permethrin, Cypermethrin, and Fenvalerate	okra fruits	GLC-ECD	73)
Cypermethrin, Deltamethrin, Fenvalerate, and Permethrin	crops, water and soils	GLC-ECD	74)
Bioresmethrin, Cismethrin, Cypermethrin, Deltamethrin, Fenpropathrin, Fenvalerate, Permethrin, Phenothrin, and Resmethrin	apples, pears, cabbage, and potatoes	GLC-ECD, HPLC: uv det.	48)
Pyrethrins, Cypermethrin, Deltamethrin, Fenvalerate, Fluvalinate, Permethrin, Resmethrin, and Tetramethrin (and about 220 other pestic.)	more than 50 different materials, mostly crops	GLC-ECD and GLC-N/FID	10)
Pyrethrins and Pip.-butox.	11 crop materials	GLC-ECD; Pip.but.: after bromination	69)
Cypermethrin, Deltamethrin, Fenpropathrin, Fenvalerate and Permethrin	17 crop materials	GLC-ECD	49)
Fenpropanate, Tetramethrin, Permethrin, Cypermethrin, Fenvalerate, Deltamethrin	9 crop materials	GLC-ECD	50)
Cypermethrin, Deltamethrin, Fenvalerate, and Permethrin	fruits and vegetables (8 differ. materials)	GLC-ECD	75)
Cypermethrin, Fenvalerate, and Permethrin	celery and animal products	GLC-ECD	36)

and found in these laboratories. In light of the multiplicity of imaginable residues this is a very difficult task, especially if besides the quantification also a reliable *identification* of seemingly or actually found residues of plant protectants must be achieved.

Multi-residue methods which allow to detect entire groups of active ingredients, therefore, facilitate the work of the quality control analyst considerably.

The situation is quite different in industrial laboratories because there the history of the sample to be analysed is usually known. On the other hand, it may be advantageous to use special, tailor-made methods in these laboratories.

Concerning the insecticidal pyrethroids respective multi-residue methods are listed in Table 2 together with references to respective publications.

2.2 Residue Data

In this chapter we will report about residue data developed in our own laboratories as well as those being available from technical literature. The major portion will cover results from analyses of plant materials. However, data on residues in soil and in materials originating from animals will also be reported. In addition, matters concerning residues resulting from the use of pyrethroids in household, hygiene and stored product protection will be discussed.

Partly, an assessment of the respective residue situation will be made in this chapter. This refers especially to the pre-harvest intervals which were calculated and proposed — based on the available residue data — by the authors of the reports or the authorities respectively.

We also refer to Chap. 3.

As far as the cyfluthrin residues are concerned, the data summarised in the respective tables were in general not further commented in the text. The reason is that in many cases final MRL-values (MRL = *M*aximum *R*esidue *L*imits) for cyfluthrin are still pending (see Table 88). As for the rest, the general remarks given in Chapter C "Final Remarks" also apply to the residue situation concerning cyfluthrin.

2.2.1 Vegetables

2.2.1.1 Brassica Leafy Vegetables

2.2.1.1.1 Cauliflower

It can be taken from Table 3 that no problems of residues on or in cauliflower are to be expected after the use of pyrethroids. S. Rai et al. [79] indicated that, given identical or even more reliable efficacy of the pyrethroids against spider mites, residue problems will more likely result if insecticidal organic phosphates such as quinalphos or trichlorfon are used.

The data summarized in the Table show that pre-harvest intervals of 15 days will be sufficient with all named pyrethroids used on cauliflower with the exception of *permethrin*. For this substance a safety interval of 3 days is indicated in one case [79], in another case, however, 22.7 or 29.4 days resp. [77]. Further investigations for clarification of the residue situation concerning permethrin on cauliflower would therefore be appropriate.

Table 3a. Residues of pyrethroids in cauliflower (Internal Residue Reports (cyfluthrin), Bayer AG, Leverkusen, FRG)

Formulation	Days after last application	Country[1]	Residue levels (mg/kg) at harvest or at days after last application	Dosage a.i. (kg/ha)
50 EC	0/4/7/11	NL	0.03/ 0.01/<0.01/<0.01	1 × 0.025
12.5 EC	0/3/5/7/10/14	D	0.10/ 0.04/ 0.03/ 0.02/ 0.02/ 0.01	3 × 0.025
12.5 EC	0/3/5/7/10/14	D	0.05/ 0.03/<0.01/<0.01/<0.01/<0.01	3 × 0.025
12.5 EC	0/3/5/7/10/14	D	0.02/<0.01/<0.01/<0.01/<0.01/<0.01	4 × 0.025
12.5 EC	0/3/5/7/10/14	D	0.07/ 0.01/<0.01/<0.01/<0.01/<0.01	3 × 0.025
12.5 EC	0/3/5/7/10/14	D	0.03/ 0.01/ 0.02/ 0.02/<0.01/<0.01	3 × 0.025
12.5 EC	0/3/5/7/10/14	D	0.02/ 0.02/ 0.01/<0.01/<0.01/<0.01	3 × 0.025
12.5 EC	0/3/5/7/10/14	D	0.13/ 0.06/ 0.02/<0.01/<0.01/<0.01	3 × 0.025
12.5 EC	0/3/5/7/10/14	D	0.06/ 0.02/<0.01/<0.01/<0.01/<0.01	3 × 0.025
50 EC	0/3/5/7/10	D	0.08/ 0.05/ 0.03/<0.02/<0.02	3 × 0.015
50 EC	0/3/5/7/10	D	0.02/<0.02/<0.02/<0.02/<0.02	3 × 0.015
50 EC	0/3/5/7/10	D	0.06/<0.02/<0.02/<0.02/<0.02	3 × 0.015
50 EC	0/3/5/7/10	D	0.09/<0.02/<0.02/<0.02/<0.02	3 × 0.015
50 EC	0/3/5/7/10	D	0.02/<0.01/<0.01/<0.01/<0.01	3 × 0.015
50 EC	0/3/5/7/10	D	0.14/ 0.10/ 0.04/ 0.03/ 0.01	3 × 0.015
50 EC	0/3/5/7/10	D	0.07/ 0.02/ 0.01/<0.01/<0.01	3 × 0.015
50 EC	0/3/5/7/10	D	0.05/ 0.02/<0.01/<0.01/<0.01	3 × 0.015
12.5 EW	0/3/5/7/10/14	D	0.04/ 0.05/ 0.03/ 0.02/ 0.02/<0.01	3 × 0.025
12.5 EW	0/3/5/7/10/14	D	0.05/ 0.02/ 0.01/ 0.01/ 0.01/<0.01	3 × 0.025
12.5 EW	0/3/5/7/10/14	D	0.04/ 0.01/ 0.01/<0.01/<0.01/<0.01	3 × 0.025
12.5 EW	0/3/5/7/10/14	D	0.05/ 0.01/<0.01/<0.01/<0.01/<0.01	3 × 0.025

[1] NL = The Netherlands; D = Federal Republic of Germany

Table 3b. Residues of pyrethroids in cauliflower (Table cont.)

Compound	Formulation	Days after last application	Country[1]	Residue levels (mg/kg) at harvest or at days after last application	Dosage a.i. (kg/ha)	Remarks (Ref.)
Cyhalothrin	n.g.	1 hr/1/3/8/11/15	IND	0.23/0.20/0.17/0.10/0.09/ND	0.03	79)
Cypermethrin	n.g.	1 hr/1/3/8/11/15	IND	0.35/0.30/0.20/0.15/0.01/0.04	0.045	79)
	25 EC	initial residues	IND	2.8	0.0075%	77 a
			4.7		0.015%	
Deltamethrin	n.g.	n.g.	n.g.	0.05	0.0125	76)
Fenpropanate (Fenpropathrin)	n.g.	1 hr/1/3/8/11/15	IND	0.28/0.22/0.16/0.09/0.06/ND	0.03	79)
Fenvalerate	n.g.	1 hr/1/3/8/11/15	IND	1.25/1.05/0.85/0.63/0.45/0.20	0.12	79)
	20 EC	initial residues	IND	2.5	0.01%	77 a
				4.4	0.02%	
	20 EC	1 hr/3/5/7/15	IND	0.86/0.60/0.44/0.30/ND 0.62/0.56/0.40/0.27/ND[b] 0.25/0.27/0.21/0.18/ND[c]	0.25	78)
Flucythrinate	n.g.	1 hr/1/3/8/11/15	IND	0.21/0.18/0.16/0.14/0.07/ND	0.03	79)
Fluvalinate	n.g.	1 hr/1/3/8/11/15	IND	0.21/0.21/0.19/0.12/0.08/ND	0.03	79)
Permethrin	n.g.	1 hr/1/3/8/11/15	IND	0.76/0.60/0.40/0.25/0.15/ND	0.09	79)
	50 EC	initial residues	IND	5.9 10.0	0.015% 0.03%	77 a

[1] IND = India; n.g. = not given; ND = not detectable (<0.01 mg/kg).

[a] Half-life (in days): 2.2 and 2.8 for cypermethrin, 2.6 and 3.5 for fenvalerate and 4.7 and 5.2 for permethrin, respectively. The waiting periods (in days) are: 8.3 and 12.8 for cypermethrin, 9.4 and 15.6 for fenvalerate and 22.7 and 29.4 for permethrin, respectively (FAO/WHO, Codex Aliment. Comm. Rep. of the Eighth Session of the CCPR; 1975).

[b] Residues after washing.

[c] Residues after cooking (the samples had been washed and after that cooked for 15 minutes in water)

Table 4a. Residues of pyrethroids in Savoy cabbage/white cabbage

Compound	Formulation	Days after last application	Country[1]	Residue levels (mg/kg) at harvest or at days after last application	Dosage a.i. (kg/ha)	Remarks (Ref.)
Cyphalothrin						see next tables
Cypermethrin	n.g.	1 hr/1/3/5/10	IND	4.80/2.64/2.20/0.94/0.35	0.015%[a]	80)
Deltamethrin	n.g.	1 hr/1/3/5/10	IND	1.35/1.00/0.08/0.05/ND[b]	0.00375%[a]	80)
	n.g.	n.g.	n.g.	<0.05	0.005 to 0.015	76)
Fenvalerate	n.g.	1 hr/1/3/5/10	IND	2.30/1.30/0.80/0.45/0.20	0.006%[a]	80)
Permethrin	n.g.	1 hr/1/3/5/10	IND	5.11/3.18/2.60/1.26/0.86	0.017%[a]	80)
	EC	0/1/7		1.34/1.07/0.38[c]	4 × 0.112	18)

[1] IND = India; n.g. = not given; [a] 8 treatments; [b] ND = not detectable; [c] mean results from the single values of the four treatments. In addition, residue values of two metabolites are given. For details see 18)

Table 4b. Residues of cyfluthrin in savoy cabbage (Table cont.) (Internal Residue Reports, Bayer AG, Leverkusen, FRG)

Formulation	Days after last application	Country[1]	Residue levels (mg/kg) at harvest or at days after last application	Dosage a.i. (kg/ha)
50 EC	0/4/7/14/21	D	0.31/0.10/0.02/ 0.02/ 0.01	3 × 0.025
50 EC	0/4/7/14/21	D	0.17/0.13/0.04/ 0.06/ 0.01	3 × 0.025
50 EC	0/4/7/14/21	D	0.30/0.08/0.01/ 0.01/ 0.01	3 × 0.025
50 EC	0/3/5/7/10/14	D	0.17/0.08/0.05/ 0.04/ 0.03/ 0.02	3 × 0.015
50 EC	0/3/5/7/10/14	D	0.33/0.19/0.15/ 0.14/ 0.07/ 0.04	3 × 0.015
50 EC	0/3/5/7/10/14	D	0.13/0.18/0.10/ 0.04/ 0.03/ <0.01	3 × 0.015
50 EC	0/3/5/7/10/14	D	0.14/0.05/0.04/ 0.02/ 0.01/ <0.01	3 × 0.015
12.5 EC	0/3/5/7/10/14	D	0.59/0.48/0.42/ 0.24/ 0.19/ 0.07	3 × 0.015
12.5 EC	0/3/5/7/10/14	D	0.31/0.17/0.09/ 0.07/ 0.07/ 0.06	3 × 0.015
12.5 EC	0/3/5/7/10/14	D	0.25/0.17/0.06/ 0.04/ 0.04/ <0.01	3 × 0.015
12.5 EC	0/3/5/7/10/14	D	0.29/0.12/0.12/ 0.07/ 0.15/ 0.15	3 × 0.015
50 EC	0/3/5/7/10	D	0.19/0.11/0.09/ 0.06/ 0.05	3 × 0.015
12.5 EC	0/3/5/7/10/14	D	0.18/0.11/0.09/ 0.05/ 0.02/ 0.01	3 × 0.025
12.5 EC	0/3/5/7/10/14	D	0.18/0.12/0.15/ 0.07/ 0.04/ 0.04	3 × 0.025
12,5 EC	0/3/5/7/10/14	D	0.22/0.15/0.09/ 0.09/ 0.05/ 0.01	3 × 0.025
12,5 EC	0/3/5/7/10/14	D	0.19/0.15/0.13/ 0.12/ 0.08/ 0.04	3 × 0.025
50 EC	0/3/5/7/10	D	0.10/0.10/ 0.04/ 0.04/ 0.03	3 × 0.015
50 EC	0/3/5/7/10	D	0.10/0.10/ 0.04/ 0.04/ 0.03	3 × 0.015
50 EC	0/3/5/7/10	D	0.52/0.30/ 0.37/ 0.30/ 0.14	3 × 0.015
50 EC	0/3/5/7/10	D	0.20/0.06/ 0.02/ 0.01/ <0.01	3 × 0.015
50 EC	0/3/5/7/10	D	0.10/0.07/ <0.01/ <0.01/ <0.01	3 × 0.015
12.5 EW	0/3/5/7/10	D	0.14/0.11/0.11/ 0.07/ 0.08/ 0.05	3 × 0.025
12.5 EW	0/3/5/7/10	D	0.10/0.08/0.07/ 0.05/ 0.04/ 0.02	3 × 0.025
12.5 EW	0/3/5/7/10	D	0.13/0.04/0.03/ 0.02/ 0.01/ <0.01	3 × 0.025
12.5 EW	0/3/5/7/10	D	0.13/0.11/0.03/ 0.03/ <0.01/ <0.01	3 × 0.025

[1] D = Federal Republik of Germany

Table 5. Lambda-cyhalothrin residues in Savoy cabbage (5% EC), Germany, 1985

Country and Year	Formulation	Rate of Application (g a.i./ha)	Volume of Spray (l/ha)	Number of Applications	Interval between Last Spraying And Sampling	Lambda-cyhalothrin levels (mg/kg)
Fed. Rep. of Germany 1985	5% EC	10	600	3	0	0.07
					1	0.06
					4	0.03
					7	0.01
					10	0.01
					14	0.04

Kinkaid, A. and Sapiets, A. 1986a "PP321: Residue Levels on Savoy Cabbage from a Trial Carried Out in Germany During 1985". ICI Plant Prot. Div. Rpt. No. M4240B

Table 6. Lambda-cyhalothrin residues in cabbages. Data obtained from cyhalothrin trials

Country and Year	Formulation	Rate of Application (g a.i./ha)	Volume of Spray (1/ha)	Number of Applications	Interval Between Last Spraying And Sampling (Days)	Lambda-cyhalothrin levels (mg/kg)
Spain 1983	5% EC	10	300	4	0	0.02
					3	0.01
					7	0.01
		20	300	4	0	0.01
					3	0.02
					7	<0.01
United Kingdom 1983 Kingdom	2.5% WP	37.5	3000	3	3	0.01, 0.03
					7	0.01, 0.004
					14	<0.004, <0.004
					7	0.01, 0.004
USA 1982	5% EC	5	225	6	0	0.04
					1	0.03
					3	0.01
					7	0.01
		10	225	6	0	0.13
					1	0.04
					3	0.02
					7	0.03
Brazil 1984	2.5% EC	3.75	400	1	8	<0.01
					15	<0.01
		5	400	1	8	<0.01
					15	<0.01
		10	400	1	8	<0.01
					15	<0.01

Canada 1983	2.5% EC	5	500	2	3	<0.01	1/
		10	500	2	3	<0.01	
Canada 1985	5% EC	10	500	1	1	<0.01, <0.01, <0.01, <0.01	1/
		20	500	1	1	<0.01, <0.01	
		10	1250	1	1	0.02, 0.01, <0.01, 0.01	1/
		20	1250	1	1	0.01, 0.02	
France 1984	5% EC	7.5	300	1	0	0.09 0.11	1/
					1	0.08 0.06	
					7	0.05 0.07	
					15	0.04 0.03	
		10	300	1	0	0.09 0.10	
					1	0.09 0.08	
					7	0.06 0.08	
					15	0.04 0.02	
		7.5	1000	1	7	0.04	
		10	1000	1	7	0.03	
		7.5	500	1	6	0.04	
		10	500	1	6	0.05	
		7.5	1000	1	7	0.06	
		10	1000	1	7	0.08	
France 1985	5% EC	7.5	500	1	0	0.24	
					1	0.19	
					3	0.11	
					7	0.06	
					14	0.03	
		10	500	1	0	0.22	
					1	0.27	
					3	0.16	
					7	0.04	
					14	0.03	

Table 6. (Continued)

Country and Year	Formulation	Rate of Application (g a.i./ha)	Volume of Spray (l/ha)	Number of Applications	Interval Between Last Spraying And Sampling (Days)	Lambda-cyhalothrin levels (mg/kg)
		7.5	300	1	0	0.09
					1	0.05
					2	0.06
					7	0.05
					15	0.01
		10	300	1	0	0.06
					1	0.10
					2	0.07
					7	0.04
					15	0.03
France 1985 cont.	5% EC	7.5	1000	1	6	0.02
					15	0.01
					104	0.01
		7.5	500	1	7	0.02
					14	0.02
					31	0.01
Malaysia 1985	2.5% EC	50	1000	– 3	5	0.01
					7	0.01
		70	1000	3	5	0.01
					7	0.01
		100	1000	3	5	0.01
					7	0.01

United Kingdom 1984	5% EC	5	400	1	0	0.04		
					3	0.05		
					7	0.02		
		10	400	1	0	0.07		
					3	0.05		
					7	0.02		
		15	400	1	0	0.08		
					3	0.08		
					7	0.04		
United Kingdom 1984 cont.	5% EC	5	260	1	0	0.01	<0.01	*2*
					3	0.02	0.01	
					7	0.02	0.01	
		10	260	1	0	0.02	0.01	
					3	0.04	0.03	
					7	0.05	0.02	
		15	260	1	0	0.06	0.03	
					3	0.05	0.06	
					7	0.09	0.05	
		5	280	1	0	0.14		*3*
		10	280	1	0	0.31		
		15	280	1	0	0.48		

The duplicate results for UK are on single samples from two different varieties on the same site sprayed at the same time. All trials involved ground application.

Swaine, H., Sapiets, A. and Burke, S. 1983e. "Cyhalothrin Residues in Cabbage from USA in 1982". ICI Plant Prot. Div. Residue Data Report no. PP563B006.

Sapiets, A. 1984h. "Cyhalotrhin Levels on Cabbage from 1983 Trials in Spain". ICI Plant Prot. Div. Rpt. No. M3790B.

Sapiets, A. 1984i. "Cyhalothrin: Residues on Cabbage from 1983 Trials in the United Kingdom". ICI Plant Prot. Div. Rpt. No. M3733B.

1/ Replicate plots; *2*/ Two varieties, one farm; *3*/ Treated and sampled while immature

All trials involved ground application.

Burke, S. and Sapiets, A. 1985. "PP321: Residue Levels in Cabbage from a 1984 Trial in Brazil". ICI Plant Prot. Div. Rpt. No. M4035B.

Culoto, B. 1985. "Study of PP321 Residue in Cabbages". ICI SOPRA Rpt. No. R5-845P, E25 834P, E26-834P (unpublished).

Kennedy, G. and Sapiets, A. 1986b. "PP321: Residue Levels on Cabbage from Trials Carried Out in Canada During 1985". ICI Plant Prot. Div. Rpt. No. M4207B.

Kennedy, G. and Sapiets, A. 1986c. "PP321: Residue Levels on Cabbage from a 1985 Trial in Malaysia". ICI Plant Prot. Div. Rpt. No. M4229B.

Massenot, F. and Culoto, B. 1986. "Study of PP321 Residues in Cabbages". ICI SOPRA Rpt. No. R 1-FP, E 38-EP (unpublished).

Sapiets, A. 1984j. "PP321: Residue Levels on Cabbage from a 1983 Trial in Canada". ICI Plant Prot. Div. Rpt. No. M3796B.

Sapiets, A. 1985e. "PP321: Residue Levels on Cabbage from 1984 Trials in The United Kingdom". ICI Plant Prot. Div. Rpt. No. M3933B.

2.2.1.1.2 Cabbage and Brussels Sprouts

As N. P. Agnihotri et al. [80] stated, the rate of dissipation of 4 pyrethroids in/on "cabbage" (variety "Drum Head") is in the order deltamethrin > cypermethrin > fenvalerate > *permethrin* (see Table 4).

As can be taken from Tables 5 and 6 (Ref. [81]), the residues of *cyhaalothrin* in/on savoy cabbage and "cabbages" decline within a few days after application to values < 0.1 mg/kg.

The residues of permethrin in/on Chinese cabbage declined to about 20% of the initial residues within 3 days (Table 10 [82]). An interesting finding is that after treatment of Chinese cabbage with permethrin the two metabolites metaphenoxybenzyl alcohol and dichlorovinyl acid were not detectable within 3 days after the application. D. A. George et al. [18], in turn, were able to detect the two metabolites in a study with permethrin in/on (head) cabbage: the corresponding zero hour residues were 1.34 mg/kg for permethrin, 1.41 mg/kg for the alcohol metabolite and 0.05 mg/kg for the acid metabolite.

The other tables need no further comments.

2.2.1.1.3 Collards (Kale)

The residue situation of pyrethroids on curly kale can be taken from Table 11.

Within the recommended pre-harvest interval of 7 days the *fenvalerate* residues were at or below 5 mg/kg.

2.2.1.2 Fruiting Vegetables

2.2.1.2.1 Paprika and Pepper

Table 12 shows some residue data of cyfluthrin in paprika (red pepper).

The residue situation of deltamethrin and fenvalerate in pepper can be taken from Table 13.

The amount of residues found was not significantly changed by doubling the applied amount of active ingredient [83].

2.2.1.2.2 Tomatoes

Numerous results are available concerning pyrethroid residues in tomatoes (see Table 14).

The data on *bioresmethrin* listed in Table 14 [84] were derived from a ^{14}C-study to investigate the fate of bioresmethrin during washing of treated tomatoes. In a laboratory test radioactively labeled parent compound was applied in form of an "EC formulation" at a rate which resulted in a surface residue of 1.25 mg/kg of tomatoes. The amount of bioresmethrin in flesh and skin was determined after washing the fruit with benzene for 2 minutes.

More than 90% of the substance was recovered if the benzene washing is performed within 24 hours after the application of the active ingredient. At the later harvest dates, up to 25% of the remaining residues were found in the skin. This amount, however, represented only less than 5% of the initially applied amount. The amount of bioresmethrin found in the flesh was not significant [84].

Table 7. Residues of cyfluthrin in white cabbage (Internal Residue Reports, Bayer AG, Leverkusen, FRG)

Formulation	Days after last application	Country[1]	Residue levels (mg/kg) at harvest or at days after last application	Dosage a.i. (kg/ha)
12.5 EC	0/3/5/7/10/14	D	0.18/ 0.16/ 0.12/ 0.06/ 0.04/<0.01	3 × 0.025
12.5 EC	0/3/5/7/10/14	D	0.08/ 0.08/ 0.06/ 0.11/ 0.04/ 0.03	3 × 0.025
12.5 EC	0/3/5/7/10/14	D	0.31/ 0.34/ 0.31/ 0.28/ 0.26/ 0.20	3 × 0.025
12.5 EC	0/3/5/7/10/14	D	0.19/ 0.05/ 0.07/ 0.06/ 0.05/ 0.07	3 × 0.025
12.5 EC	0/3/5/7/10/14	D	0.05/ 0.01/<0.01/<0.01/<0.01/<0.01	3 × 0.025
12.5 EC	0/3/5/7/10/14	D	0.15/ 0.07/ 0.06/ 0.08/ 0.02/ 0.04	3 × 0.025
50 EC	0/3/5/7/10	D	0.28/ 0.11/ 0.07/ 0.04	3 × 0.015
50 EC	0/3/5/7/10	D	<0.02/<0.02/<0.02/<0.02/<0.02/<0.02	3 × 0.015

[1] D = Federal Republik of Germany

Table 8. Residues of cyfluthrin in red cabbage (Internal Residue Reports, Bayer AG, Leverkusen, FRG)

Formulation	Days after last application	Country[1]	Residue levels (mg/kg) at harvest or at days after last application	Dosage a.i. (kg/ha)
12.5 EC	0/3/5/7/10/14	D	0.2/<0.01/<0.01/<0.01/<0.01/<0.01	3 × 0.025
12.5 EC	0/3/5/7/10/14	D	0.01/<0.01/<0.01/<0.01/<0.01/<0.01	3 × 0.025
50 EC	0/3/5/7/10		0.04o 0.2/ 0.02/ 0.03/ 0.03	3 × 0.015
50 EC	0/3/5/7/10	D	<0.02/<0.02/<0.02/<0.02/<0.02	3 × 0.015

[1] D = Federal Republik of Germany

Table 9a. Residues of cyfluthrin in Brussels sprouts (Internal Residue Reports, Bayer AG, Leverkusen, FRG)

Formulation	Days after last application	Country[1]	Residue levels (mg/kg) at harvest or at days after last application	Dosage a.i. (kg/ha)
12.5 EC	0/3/5/7/10/14	D	0.06/0.09/0.05/0.06/0.07/0.06	3 × 0.025
12.5 EC	0/3/5/7/10/14	D	0.23/0.17/0.16/0.13/0.06/0.10	3 × 0.025
12.5 EC	0/3/5/7/10/14	D	0.02/0.03/0.03/0.02/0.02/0.01	3 × 0.025
12.5 EC	0/3/5/7/10/14	D	0.05/0.06/0.06/0.05/0.05/0.05	3 × 0.025
12.5 EC	0/3/5/7/10/14	D	0.02/0.02/0.03/0.11/0.06/0.09	3 × 0.025
12.5 EC	0/3/5/7/10/14	D	0.03/0.03/0.05/0.05/0.05/0.05	3 × 0.025
12.5 EC	0/3/5/7/10/14	D	0.06/0.09/0.08/0.08/0.03/0.06	3 × 0.025
12.5 EC	0/3/5/7/10/14	D	0.18/0.01/0.02/0.05/0.01/0.01	3 × 0.025
12.5 EW	0/3/5/7/10/14	D	0.04/0.02/0.02/0.02/0.03/0.02	2 × 0.025/1 × 0.038
12.5 EW	0/3/5/7/10/14	D	0.07/0.11/0.09/0.09/0.11/0.06	3 × 0.038
12.5 EW	0/3/5/7/10/14	D	0.04/0.02/0.02/0.03/0.02/0.01	2 × 0.025/1 × 0.038
12.5 EW	0/3/5/7/10/14	D	0.13/0.16/0.14/0.10/0.15/0.11	3 × 0.038

[1] D = Federal Republik of Germany

Table 9b. Residues of deltamethrin in Brussels sprouts

Compound	Formulation	Days after last application	Country	Residue levels (mg/kg at harvest or at days after last application	Dosage a.i. (kg/ha)	Remarks (refer.)
Deltamethrin	n.g.	n.g.	n.g.	"no residues detected"	6 × 0.0125	(76)

n.g. = not given

Table 10. Residues of pyrethroids in Chinese cabbage

Compound	Formulation	Days after last application	Country	Residue levels (mg/kg) at harvest or at days after last application	Dosage a.i. (kg/ha)	Remarks (Ref.)
Deltamethrin	n.g.	6	n.g.	0.03	0.05	[76]
Permethrin	EC	0 hr/1/3	USA	0.29/0.29/0.05[a]	0.112	[82]
	EC	0 hr/1/3	USA	0.94/0.32/0.20[a]	0.224	[82]

[a] 8 treatments at one-week intervals.
Additionally the samples were analysed for residues of the two metabolites metaphenoxybenzyl alcohol and dichlorovinyl acid. Result: All the values were <0.05 mg/kg.
n.g. = not given

Table 11. Residues of pyrethroids in collards (kale)

Compound	Formulation	Days after last application	Country	Residue levels (mg/kg) at harvest or at days after last application	Dosage a.i. (kg/ha)	Remarks (Ref.)
Deltamethrin	n.g.	3	n.g.	0.007	0.0125	[76]
Fenvalerate	30 EC	3/7/10	USA (Georgia)	8.7/5.5/2.4 (average)	4 × 0.224	[81]
	30 EC	3/7/10	USA (Virginia)	4.4/1.6/0.34 (average)[a]	5 × 0.224	[81]
	30 EC	3/7/10	USA (Virginia)	6.8/3.7/2.1[a]	5 × 0.448	[81]
	30 EC	3/7/10	USA (Maryland)	3.4/2.7/1.5 (average)[a]	5 × 0.224	[81]
	30 EC	3/7/10	USA (Maryland)	4.9/4.8/2.3[a]	5 × 0.448	[81]

[a] Additionally most of the samples were analysed for residues of the metabolite SD 54597 (decarboxylated active ingredient). The values found ranged between <0.08 and 0.98 mg/kg (for further details see [81]).
n.g. = not given

Table 12. Residues of cyfluthrin in paprika (red pepper) (Internal Residue Reports, Bayer AG, Leverkusen, FRG)

Formulation	Days after last application	Country[1]	Residue levels (mg/kg) at harvest or at days after last application	Dosage a.i. (kg/ha)
50 SL	0/1/3/4/7	E	0.05/0.03/0.02/0.02/0.01	3 × 0.025
50 SL	0/1/3/4/7	E	0.05/0.03/0.03/0.04/0.04	3 × 0.025
25 EC	0/3/8/14	E	0.04/0.03/0.03/ < 0.02	3 × 0.05
25 EC	0/3/8/14	E	0.04/0.07/0.03/ < 0.02	3 × 0.05

[1] E= Spain

M. D. Awasthi [89)] studied the influence of urea upon the persistence of *various pyrethroids* in/on tomatoes. The results are also summarized in Table 14. Urea provides "foliar nutrition" and could increase the speed of degradation and/or favour the washing off of the active ingredient.

The studies of Awasthi show that urea does not exert any influence upon the persistence of the active ingredient. Table 14 also allows for a conclusion to be drawn on the relative persistenceof different pyrethroids on tomatoes: *cypermethrin > fenvalerate > permethrin > deltamethrin.*

The *deltamethrin residues* in tomatoes reported under [90)] were taken from the official testing programme of pesticides, Finland (1981).

In greenhouse experiments permethrin residues shown under [85)] and [86)] were obtained. In these experiments the parent compound turned out to be relatively stable on greenhouse tomatoes. According to [85)] a safety interval of 8 days is proposed for greenhouse tomatoes (treatment at a 0.0075% concentration).

According to [86)] the differences in the residue levels between the cis-and the trans-isomer of permethrin on greenhouse tomatoes are not significant.

According to [87)] permethrin shows the longest persistence and consequently likely also the greatest effectiveness on tomatoes in comparison to the pyrethroids *fenpropanate and fenvalerate A* (= isomer mixtures of ± S; alpha S and R; alpha R resp.).

The residue levels of deltamethrin and permethrin in/on greenhouse tomatoes summarized in Table 14 under [88)] are regarded by the authors to be "quite safe" considering the tolerance level of 0.50 mg/kg for deltamethrin and 1.00 mg/kg for permethrin.

In a collaborative study with 5 laboratories participating, residues of *fenvalerate* were determined in processed tomato products and by-products and chopped whole tomatoes [54)].

The data listed in Table 14 under [78)] originate from a processing study. The initial deposits (0.85 mg/kg) were reduced by approx. 50% by washing with water. In case of samples taken later, only 3–7% were washed off. Cooking of the washed tomatoes in water for 15 mins. resulted in losses of approx. 88% of that amount of active ingredient which was present on the fruit immediately after application of the active ingredient, while only approx. 70% of the fenvalerate residues were eliminated from samples taken later by cooking the fruit. As the residues were < 1 mg/kg in all analysed samples, the use of fenvalerate on tomatoes can be con-sidered as "safe" (WHO Techn. Rep. Ser. No. 525 (1973)).

Table 13. Residues of pyrethroids in pepper

Compound	Formulation	Days after last application	Country[1]	Residue levels (mg/kg at harvest or at days after last application	Dosage a.i. (kg/ha)	Remarks (Ref.)
Deltamethrin	n.g.	4	n.g.	<0.03	0.0125	[76]
Fenvalerate	"20%"	0/7/14	ISR	0.15/0.12/0.06	0.15	[83]
		0/7/14	ISR	0.08/0.12/0.11	0.30	[83]

[1] ISR = Israel
n.g. = not given

Table 14a. Residues of cyfluthrin in tomatoes (Internal Residue Reports, Bayer AG, Leverkusen, FRG)

Formulation	Days after last application	Country[1]	Residue levels (mg/kg) at harvest or at days after last application	Dosage a.i. (kg/ha)
50 SL	0/1/3/4/7	D	0.05/ 0.05/ 0.03/ 0.03/ 0.02	3 × 0.03
50 SL	0/1/3/4/7	D	0.02/ 0.01/ 0.02/ 0.02/ 0.02	2 × 0.22/1 × 0.03
50 SL	0/1/3/4/7	I	0.11/ 0.06/ 0.06/ 0.04/ 0.03	3 × 0.038
50 SL	0/1/3/4/7	E	0.05/ 0.02/ 0.02/ 0.02/ 0.01	3 × 0.025
50 SL	0/1/3/4/7	E	0.06/ 0.03/ 0.03/ 0.03/ 0.01	3 × 0.025
50 SL	0/1/3/4/7	I	0.02/ 0.02/ 0.01/ <0.01/ <0.01	3 × 0.038
50 EC	0/1/3/5	NL	0.03/ 0.05/ 0.03/ 0.03	1 × 0.075
25 EC	0/3/8/14	E	<0.02/ <0.02/ <0.02/ <0.02	3 × 0.05
25 EC	0/3/8/14	E	0.02/ 0.02/ <0.02/ <0.02	3 × 0.05
50 EC	0/3/5/7/10	D	<0.01/ <0.01/ <0.01/ <0.01/ <0.01	0.015/0.023/0.03
50 EC	0/3/5/7/10	D	0.03/ 0.01/ 0.01/ <0.01/ <0.01	3 × 0.023
50 EC	0/3/5/7	D	0.03/ 0.02/ 0.01/ 0.01	3 × 0.023
50 EC	0/3/5/7/10	D	0.04/ 0.04/ 0.02/ 0.02/ 0.01	2 × 0.023/1 × 0.03
50 EC	0/3/5/7/10	D	0.03/ 0.03/ 0.03/ 0.03/ <0.01	2 × 0.023/1 × 0.03
50 EC	0/3/5/7/10	D	0.05/ 0.04/ 0.03/ <0.01/ <0.01	0.015/0.023/0.03
50 EC	0/3/5/7/10	D	0.04/ 0.04/ 0.02/ 0.01/ <0.01	3 × 0.023
50 EC	0/3/5/7/10	D	0.12/ 0.12/ 0.12/ 0.10/ 0.10	3 × 0.015
50 EC	14	Mar	0.04	6 × 0.025

[1] D = Federal Republic of Germany; I = Italy; E = Spain; NL = The Netherlands; Mar = Marocco

Table 14b. Residues of pyrethroids in tomatoes

Compound	Formulation	Days after last application	Country[1]	Residue levels (mg/kg) at harvest or at days after last application	Dosage a.i. (kg/ha)	Remarks (Ref.)
Bioresmethrin	EC	1 hr/6 hrs/12 hrs 1/2/3 1 hr/6 hrs/12 hrs 1/2/3	n.g.	*flesh*: 0.2/0.2/0.2[a] 0.2/0.2/0.2 *skin*: 1.75/2.45/3.2[a] 1.95/4.65/4.65	1.25 mg/kg tomatoes	[84]
Cypermethrin	n.g.	1hr/3/7/10/15/21	IND	1.02/0.74/0.43/0.22/ND/— 1.02/0.71/0.38/0.18/ND/—[b]	0.1 0.1 + urea	[89]
Deltamethrin	n.g.	2–3	n.g.	0.02	0.0125	[76]
	n.g.	0/1/2/4/7/14/21	I	0.08/0.06/0.06/0.08/0.13/0.07/0.06 0.16/0.16/0.18/0.14/0.16/0.08/0.11	3× 0.07 3 × 0.14	[88] [88]
	n.g. n.g.	1 hr/3/7/10/15/21 1 hr/3/7/10/15/21	IND IND	0.41/0.16/0.03/ND/—/— 0.41/0.09/ND/—/—/—[c]	0.02 0.02 + urea	[89] [89]
	n.g.	1/5/8	SF	<0.05/<0.05/<0.05	1 mg/plant, plastic-greenhouse	[90]
Fenpropathrin (Fenpropanate)	n.g.	30	CAN	0.03	0.14	[87]
Fenvalerate	n.g.	30	CAN	0.01	0.14	[87]
	n.g.	1 hr/3/7/10/15/21 1 hr/3/7/10/15/21	IND IND	0.93/0.71/0.43/0.21/0.08/ND 0.93/0.68/0.41/0.18/0.03/ND[d]	0.1 0.1 + urea	[89] [89]
	2.4 EC (2.4 lb a.i.) per gal)	1	USA	0.26 chopped whole tomatoes <0.03 canned tomato quarters <0.01–0.04 tomato juice 0.12 tomato paste 0.12 tomato paste 1.9 tomato seeds plus skins	13 × 0.112	[54 d)]
	0.05% EC	1 hr/3/5/7/15/30	IND	0.85/0.76/0.67/0.52/0.30/ND[e] 0.41/0.71/0.63/0.49/0.29/ND[f] 0.1/0.18/0.15/0.13/0.29/ND[g]	0.25 (0.05% EC, 500 l/ha)	[78]

Table 14b. (Continued)

Compound	Formulation	Days after last application	Country[1]	Residue levels (mg/kg) at harvest or at days after last application	Dosage a.i. (kg/ha)	Remarks (Ref.)
Permethrin	"Ambush", 0.03%	0/6/8/16/20/25	BUL	0.8/0.2/0.1/0.08/0.045/0.04	n.g.	[85]
	n.g.	1/3/7/10, 2nd spray. 11/13/17/22/30	NZ	0.29/0.25/0.27/0.23 0.44/0.31/0.29/0.33/0.30	5 g a.i./ 100 l water to run-off	[86]
		1/3/7/10, 2nd spray. 11/13/17/22/30		0.03/0.08/0.07/0.03 0.03/0.03/0.09/0.03/0.03	1 g a.i./ 1000 plants	
	n.g.	30	CAN	0.12	0.14	[87]
	n.g.	0/1/2/4/7/14/21	I	0.10/0.07/0.12/0.15/0.15/0.10/0.05	3× 0.4	[88]
				0.14/0.12/0.20/0.26/0.20/0.12/0.09	3 × 0.8	[88]
	n.g.	1 hr/3/7/10/15/21	IND	1.38/0.93/0.51/0.22/0.03/ND	0.15	[89]
				1.38/0.88/0.50/0.22/ND/—[h]	0.15+ urea	[89]
Permethrin	unknown[i]	unknown[i]	DK	0.16/0.18 in 1982 0.07 in 1983	[i]	[91]

[1] BUL = Bulgaria; CAN = Canada; DK = Denmark; I = Italia; IND = India; NZ = New Zealand; SF = Finland; n.g. = not given

[a] The aim of the study was to determine the fate of bioresmethrin in and on tomatoes. The initial deposit of ^{14}C-bioresmethrin calculated for the fruit was 1.25 mg/kg. The results are expressed as a percentage of the total radioactivity applied. For further details see text.

[b] For foliar nutrition besides the a.i. a 1% urea solution was applied weekly.

[c] For foliar nutrition besides the a.i. a 1% urea solution was applied weekly.

[d] cooperative study; all values given are mean values from 5 laboratories

[e] unprocessed; [f] after washing with water; [g] after washing and cooking for 15 min. in water

[h] For foliar nutrition besides the a.i. a 1% urea solution was applied weekly.

[i] results of the Danish market control

Table 15a. Residues of cyfluthrin in cucumbers (Internal Residue Reports, Bayer AG, Leverkusen, FRG)

Formulation	Days after last application	Country[1]	Residue levels (mg/kg) at harvest or at days after last application	Dosage
50 EC	0/1/3/5	N	0.02/0.02/0.01/ <0.01	1 × 0.075
25 EC	0/3/8/14	E	0.03/0.03/ <0.02/ <0.02	1 × 0.05
25 EC	0/3/8/14	E	0.03/0.03/ <0.02/ <0.02	1 × 0.05

[1] NL = The Netherlands; E = Spain

The results reported in Table 14 under [91] were obtained on import tomatoes in the course of food quality controls in Denmark. The valid MRL in Denmark corresponds to the FAO/WHO guidelines and is 2 mg/kg for permethrin.

2.2.1.2.3 *Cucumbers*

Pyrethroid residues in cucumbers can be taken from Table 15. As far as *bioresmethrin* residues in this crop are concerned, it is noticeable that this substance is substantially degraded within one hour after application while this process progresses more slowly thereafter (24% remaining after one hour, 10% after 72 hours). A simple washing in the intact fruit with benzene removed at least 85% of the residual a.i. at any time up to 3 days after treatment [84]. The amount of active ingredient found in the flesh is negligible (<0.1% of concentration applied).

The residues of *deltamethrin* in cucumbers reported under Ref [90] were taken from the official testing programme of pesticides, Finland (1981).

Greenhouse trials yielded the results listed under [85] and [86]. The pre-harvest interval after treatment of cucumbers with permethrin at an a.i. concentration of 0.0075% is indicated to be 5 days for the crop grown under glass [85].

Concerning the permethrin residues on/in cucumbers reported in Table 15 under [86] it should be added that the differences between the residues of cis-versus trans-isomer on greenhouse cucumbers are not significant.

2.2.1.2.4 *Okra*

Residue levels of pyrethroids in okra are summarized in Table 16. Concerning the residues of cypermethrin, fenvalerate and permethrin listed under [73], the rates of dissipation of fenvalerate were 46.8%, 74.5% and 87.2% on the first, third and seventh day, respectively, and they were 50.0%, 72.3% and 88.6% for permethrin, respectively. The differences between the values after the first and after the second application were low.

Due to the low rate of application the initial residues of cypermethrin were correspondingly low: 1.1 mg/kg and 1.3 mg/kg. The rate of loss of cypermethrin was 5.4% to 12.5% lower than that of fenvalerate and permethrin.

A cooking process was very effective with regard to the removal of the pyrethroids. Between 50 and 80% of the residues were removed upon cooking which was done after washing with water.

Table 15b. Residues of pyrethroids in cucumbers

Compound	Days after last	Days after last application	Country[1]	Residue levels (mg/kg) at harvest or at days after last application	Dosage a.i. (kg/ha)	Remarks (Ref.)
Bioresmethrin	EC	1 hr/6 hrs/12 hrs 1/2/3	n.g.	*flesh:* 0.1/0.1/0.1[a]	1 mg/kg cucumbers	[84]
		1 hr/6 hrs/12 hrs 1/2/3		*skin:* 0.8/1.6/0.7[a] 1.4/0.8/1.2		
Deltamethrin	n.g.	2	n.g.	<0.01 (whole fruits)	n.g.	[76]
	n.g.	1/5/8	SF	<0.05/<0.05/<0.05	2 mg/plant plastic-greenhouse	[90]
Permethrin	Ambush, 0.03%	1/2/4/6	BUL	0.56/0.12/0.10/0.02	n.g.	[85]
	n.g.	1/3/7/10, 2nd spray, 11/13/17/22/30	NZ	0.13/0.05/0.02/0.03 0.18/0.04/0.02/0.02/ND	5 g a.i./ 100 l water to run-off	[86]
		1/3/7/10, 2nd spray, 11/13/17/22/30		ND/ND/ND/ND 0.03/0.02/0.02/0.01/0.02	2 g a.i./ 1000 plants	
Cypermethrin	n.g.	0/1/3/7, 2nd appl., 0/1/3/7	IND	1.1/0.7/0.5/0.2 1.0/0.8/0.6/0.2	0.04	[73]
		0/1/3/7, 2nd appl., 0/1/3/7		1.3/1.0/0.8/0.3 1.2/1.0/0.8/0.3	0.06	
	25 EC	1 hr/3/7/10	IND	0.92/0.74/0.16/ND	0.0375	[92]
	25 EC	1 hr/1/3/5/7 1 hr/1/3/5/7/10	IND	1.16/0.87/0.60/0.23/ND 1.56/1.13/0.81/0.35/0.11/ND	0.05 0.075	[93]
Deltamethrin	2.8 EC	1 hr/1/3/5 1 hr/1/3/5	IND	0.28/0.15/0.041/ND 0.35/0.17/0.044/ND	0.010 0.015	[93]

Fenvalerate	n.g.	0/1/3/7, 2nd appl., 0/1/3/7	IND	4.7/2.5/1.2/0.6 4.9/2.5/1.1/0.5	0.50	73)
	20 EC	1 hr/3/5/7/15	IND	4.0/2.8/1.6/0.8/ND (unprocessed) 1.6/2.4/1.4/0.7/ND (after washing) 0.6/1.6/1.0/0.5/ND (after cooking)	0.025	78)
	n.g.	1 hr/3/7/10	IND	2.15/1.80/0.42/ND	0.10	92)
	20 EC	1 hr/1/3/5/7/10 1 hr/1/3/5/7/10	IND	1.96/1.25/0.82/0.31/0.19/ND 3.27/2.15/1.40/0.52/0.22/ND	0.075 0.10	93)
Permethrin	n.g.	0/1/3/7, 2nd appl., 0/1/3/7	IND	4.4/2.2/1.0/0.5 4.4/2.1/1.0/0.4	0.50	73)
	25 EC	1 hr/3/7/10	IND	1.75/0.98/0.24/ND	0.075	92)
	50 EC	1 hr/1/3/5/7/10 1 hr/1/3/5/7/10	IND	1.59/1.17/0.78/0.27/0.18/ND 2.98/2.06/1.30/0.61/0.23/ND	0.075 0.10	93)

[1] n.g. = not given; ND = not detectable; SF = Finland; IND = India; BUL = Bulgaria; NZ = New Zealand

[a] The aim of the study was to determine the fate of bioresmethrin in and on cucumbers. The initial deposit of ^{14}C-bioresmethrin on the fruit was 2 mg/kg. The results are expressed as a percentage of the total radioactivity applied. For further details see text

According to the authros [73] the residue situation is safe if the safety intervals of 3 days for permethrin and cypermethrin as well as 7 days for fenvalerate are observed.

The tolerance limits are indicated to be 0.5 mg/kg for permethrin and cypermethrin and 0.2 mg/kg for fenvalerate in or on okra (Agric. Chem. News and Reports, No. 104, p. 13 and No. 107, p. 8 (1978).

The following remarks should be made concerning the fenvalerate residues on okra reported under [78]: 30% of the initial residues of 4 mg/kg are lost after 3 days and 80% after 7 days. This is also attributable to the heavy growth rate of the plants (decline more due to dilution by increase in weight and less due to degradation). In this case a safety interval of 7 days is indicated for okra after the use of fenvalerate.

From the study under [92] it can be concluded that the residues of permethrin and cypermethrin fall below the MRL values of 0.5 mg/kg after a pre-harvest interval of 7 days while the fenvalerate residues fall below the MRL value of 0.2 mg/kg after a pre-harvest interval of 10 days.

The persistence on okra according to [93] is 7 days for fenvalerate and permethrin, 5 days for cypermethrin and 3 days for deltamethrin. The safety intervals are 7, 4, 3 and 0 days for fenvalerate, permethrin, cypermethrin and deltamethrin respectively.

2.2.1.3 Leafy Vegetables

2.2.1.3.1 Lettuce

The residue situation of pyrethroids on head lettuce can be taken from Table 17.

Table 17 a. Residues of cyfluthrin in head lettuce (Internal Residue Reports, Bayer AG, Leverkusen, FRG)

Formulation	Days after last application	Country[1]	Residue levels (mg/kg) at harvest or at days after last application	Dosage a.i. (kg/ha)
50 EC	0/4/7/14/21	D	0.43/0.23/0.10/0.05/0.01	3 × 0.025
50 EC	0/4/7/14/21	D	0.59/0.36/0.13/0.08/0.05	3 × 0.025
50 EC	0/4/7/14/21	D	1.20/0.55/0.11/0.03/0.01	3 × 0.025
50 SL	0/4/7/14/21	D	0.19/0.12/ 0.07/ 0.02/<0.01	3 × 0.025
50 SL	0/4/7/14/21	D	0.35/0.19/ 0.05/<0.01/ 0.01	3 × 0.025
50 SL	0/4/7/14/21	D	0.11/0.03/<0.01/<0.01/<0.01	3 × 0.025
12.5 EC	0/3/5/7/10/14	D	1.60/0.52/0.35/0.34/0.15/0.13	3 × 0.025
12.5 EC	0/3/5/7/10/14	D	0.53/0.30/0.19/0.10/0.09/0.02	3 × 0.025
12.5 EC	0/3/5/7/10/14	D	0.75/0.46/0.40/0.24/0.13/0.13	3 × 0.025
12.5 EC	0/3/5/7/10/14	D	0.28/0.02/0.02/0.02/0.01/0.01	3 × 0.025
12.5 EC	0/3/5/7/10/14	D	0.29/0.08/0.07/0.03/0.01/0.01	3 × 0.025
12.5 EC	0/3/5/7/10/14	D	0.32/0.21/0.08/0.07/0.05/0.02	3 × 0.025
12.5 EC	0/3/5/7/10/14	D	0.73/0.32/0.29/0.19/0.13/0.05	3 × 0.025
12.5 EC	0/3/5/7/10/14	D	1.00/0.39/0.24/0.19/0.06/0.03	3 × 0.025
12.5 EW	0/3/5/7/10/14	D	0.61/0.35/ 0.17/ 0.06/<0.01/<0.01	3 × 0.025
12.5 EW	0/3/5/7/10/14	D	0.68/0.15/ 0.11/ 0.09/ 0.06/ 0.06	3 × 0.025
12.5 EW	0/3/5/7/10/14	D	0.43/0.09/ 0.08/ 0.03/<0.01/<0.01	3 × 0.025
12.5 EW	0/3/5/7/10/14	D	0.32/0.09/<0.01/<0.01/<0.01/<0.01	3 × 0.025

[1] D = Federal Republik of Germany

Table 17b. Residues of pyrethroids in lettuce

Compound	Formulation	Days after last application	Country[1]	Residue levels (mg/kg) at harvest or at days after last application	Dosage a.i. (kg/ha)	Remarks (Ref.)
Cypermethrin	"Ripcord"	0/1/3/8/10/14 (1979) 0/1/3/7/10/14 (1980)	CAN	0.24/0.20/0.072/0.004/<0.001/ND 0.50/0.50/0.43/0.097/0.035/0.011	3 × 0.08 3 × 0.08	97)
Deltamethrin	2.5 EC	12/19/26	B	0.23/0.12/0.10[a]	0.0125	96)
Fenvalerate	"Belmark"	0/1/3/7/10/14	CAN	0.43/0.48/0.21/0.15/0.069/0.054	3 × 0.08	97)
	n.g.	30/120/1 year[b]	USA	0.03–0.01	2.2	98)
Permethrin	Granulate	at harvest[c]	CAN	<0.005 in the heads —0.030 in the roots 0.057 in the roots	0.56 or 1.12 0.56 1.12	94)
	n.g.	30/60/120	UK	0.02(0.08[d])/0.03(0.08)/0.03/(0.03)	2.0	95)
	25 EC	12/19/26 3/15/27 3/15/27 4/10	B	3.1/2.0/1.5[e] 0.61/0.56/0.22 1.2/1.4/0.47 0.17/0.04	0.25 2 × 0.025 2 × 0.050 0.025	96)
Permethrin	"Ambush"	0/1/3/8/10/14	CAN	0.16/0.049/0.006/0.006/0.001/ND	3 × 0.08	97)
	n.g.	1	USA	0.4 (cis-permethrin) 0.6 (trans-permethrin)	0.22	62)
		1/2/ 1/2/		1.0/<0.2 (cis-permethrin) 1.6/0.6 (trans-permethrin)	0.34	
	unknown	unknown	DK	0.01 and 0.14 (in 1982)[f] 0.51 (in 1983)	unknown	91)

[1] CAN = Canada; UK = United Kingdom; DK = Denmark; B = Belgium; n.g. = not given; ND = not detectable;
[a] mean values from two experiments;
[b] soil treatment, planting 30 days, 120 days and 1 year after treatment;
[c] soil application; [d] in-door experiment using flower-pots; values in parentheses after ^{14}C-acid-labeling;
[e] mean values from two replicate experiments;
[f] results of the Danish market control

Concerning the residues of cypermethrin, fenvalerate and permethrin reported under[97], applications were made 3 times at rates of 80 g a.i./ha each at weekly intervals. The following residue situation resulted: the cypermethrin values declined to <0.1 mg/kg within 3–7 days, the fenvalerate residues reached this level after 10–14 days and the permethrin residues were <0.1 mg/kg already one day after the last application. Levels of <0.1 mg/kg can be considered as negligibly low. In this study [97] the greatest persistence was observed with fenvalerate.

The values reported under [96] are based on applications at the normal prescribed use rates. Even if the application of the active ingredients was made a few days before harvest, the residues of permethrin and deltamethrin resulting at the time of harvest were <1 mg/kg. The MRL value for permethrin is 1 mg/kg, that for deltamethrin is 0.2 mg/kg [96]. As deltamethrin is more persistent than permethrin, this value appears to be unrealistic because this substance is considered to be useful for treatments close to harvest. If higher rates of application are used or two applications within two weeks prior to harvest are made, residue levels are likely to exceed the MRL values.

The data reported under [98] are based on a study concerning the uptake of fenvalerate by head lettuce from the soil. The lettuce was planted 30 days, 120 days and 1 year after the application of the active ingredient to the soil (2.2 kg a.i./ha!).

After application of a granular formulation to the soil, permethrin residues were only detected in the root system of the plants [94], while the edible portions contained residues of <0.005 mg/kg. The head lettuce was sown immediately after treatment and the harvest samples were analysed.

Head lettuce was sown into ^{14}C-permethrin-treated soil 30, 60 and 120 days after application [95]. Flower pots were used for this test. The ^{14}C-contents in the mature crops were determined. The residues were calculated as permethrin. The values in parentheses resulted after ^{14}C-acid-labeling, the other values after ^{14}C-alcohol-labeling. The values in parentheses are in general higher than the other values which indicates that metabolites deriving from the acid moiety constitute a larger portion of the residues taken up from the soil. The longer the time-span between treatment and sowing, the more rapidly the residues decline.

The relatively low residue levels reported under [62] are probably due to high temperatures and intense incident radiation as well as dilution of the residues owing to rapid plant growth.

In the course of the official Danish food quality control during 1982 and 1983, permethrin residues were determined in two head lettuce samples in 1982 and in one sample in 1983. They were far below the Danish MRL value (10 mg/kg). Regrettably the total number of food samples analysed for pyrethroid residues was not reported [91].

2.2.1.3.2 Spinach

Pyrethroid residues in spinach are summarized in Table 18. Concerning the fenvalerate residues reported under [78], the substance persisted in spinach for approx. 30 days. Only on the day of application, washing with cold water resulted in a loss of approx. 50% of the initial residue. Three days after application only 3–13% of the active ingredient could be washed off. By cooking the sample taken on the day of application, approx. 80% of the active ingredient were removed. Only 70% of the residue were removed by cooking of samples taken later.

Table 18. Residues of pyrethroids in spinach

Compound	Formulation	Days after last application	Country[1]	Residue levels (mg/kg) at harvest or at days after last application	Dosage a.i. (kg/ha)	Remarks (Ref.)
Fenvalerate	EC	0/15	IND	9.5/2.8	0.25	78)
Permethrin	EC	0/1/3/7	USA	1.61/1.15/0.79/0.29	0.056	82)
		0/1/3/7		2.17/1.73/1.02/0.44	0.112	
		0/1/3/7		1.42/1.39/0.67/0.26[a]	0.056	
		0/1/3/7		2.11/1.64/0.90/0.52[a]	0.112	
		0/1/3/7		1.26/0.94/0.68/0.22[b]	0.056	
		0/1/3/7		1.82/1.49/0.89/0.39[b]	0.112	

[1] IND = India;
[a] values belonging to the respective alcohol metabolite: metaphenoxybenzyl alcohol;
[b] values belonging to the respective dichlorovinyl acid metabolite

Table 19. Residues of pyrethroids in onions

Compound	Formulation	Days after last application	Country[1]	Residue levels (mg/kg) at harvest or at days after last application	Dosage a.i. (kg/ha)	Remarks (Ref.)
Cypermethrin	40 EC	0/1/3/7/10	CAN	0.17/0.13/0.17/0.09/0.05 (1979)	0.08	[99]
		0/1/3/7/10		0.05/0.04/0.03/0.02/0.02 (1980)	0.08	
	25 EC	1 hr/3/7/15/harv.[a]	IND	0.12/0.05/0.02/0.01/ND	0.01%-spray conc.	[92]
		1 hr/3/7/15/harv.[a]		0.20/0.11/0.03/0.02/ND	0.02%-spray conc.	
Deltamethrin	n.g.	1 hr/3/7/15/harv.[a]	IND	0.08/0.06/0.03/0.05/ND	0.002%-spray con.	[92]
		1 hr/3/7/15/harv.[a]		0.15/0.11/0.04/0.02/ND	0.004%-spray conc.	
Fenvalerate	n.g.	1 hr/3/7/15/harv.[a]	IND	0.11/0.07/0.03/0.01/ND	0.01%-spray conc.	[92]
		1 hr/3/7/15/harv.[a]		0.20/0.11/0.04/0.02/ND	0.02%-spray conc.	
Permethrin	50 EC	0/1/3/7/10	CAN	0.15/0.09/0.06/0.04/0.07	0.08	[99]
		0/1/3/7/10		0.02/0.02/0.01/0.005/0.004	0.08	
	25 EC	1 hr/3/7/15/harv.[a]	IND	0.24/0.11/0.05/0.02/ND	0.01%-spray conc.	[92]
		1 hr/3/7/15/harv.[a]		0.42/0.23/0.08/0.03/ND	0.02%-spray conc.	
	Granulate	n.g.	CAN	0.030 in the roots	0.56	[94]
				0.060 in the roots	1.12	

[1] CAN = Canada; IND = India; n.g. = not given; ND = not detectable (<0.02 mg/kg); [a] onion bulb at harvest

The MRL value for fenvalerate on spinach is 1 mg/kg [78]. As in this study residues of 2.8 mg/kg were still determined after a safety interval of 15 days, this safety interval is obviously not sufficient when applying 250 g of fenvalerate/ha to spinach.

The following is yet to be said concerning the results reported under [82]: the permethrin dissipation is 50 % within 3 days after treatment and approx. 80 % within 7 days. The acid metabolite peaked 3 days after the treatment and then declined.

2.2.1.4 Bulb Vegetables

2.2.1.4.1 Onions and Leeks

The residue situation in onions is presented in Table 19. The values presented under [99], which cover cypermethrin and permethrin and were obtained in the years 1979 and 1980, refer to dry onions from spraying, after pulling and windrowing but before topping.

In 1979 the cypermethrin levels declined within 7 days to residues 0.1 mg/kg, while this level was reached after permethrin treatment already within one day. In the experiment repeated in 1980, levels 0.1 mg/kg were reached already immediately after treatment with both active ingredients.

The values reported under [92] show that deltamethrin persists longer than permethrin, cypermethrin and fenvalerate. In all three cases the residues fell below the respective MRL values [92] already within one hour after application.

[94] shows the residue situation in onions after treatment of the soil with a granular formulation. The active ingredient was not translocated into the edible portions of the onions but was translocated into the root system.

The residue situation of cyfluthrin in leeks is presented in Table 30.

2.2.1.5 Legume Vegetables

2.2.1.5.1 Beans

Residues of pyrethroids in beans can be taken from Table 21. The data reported in this Table under [100] refer to pods of French beans and cover the active ingredients cypermethrin. The respective two treatments each were made at the flowering stage and at pod development. The MRL values (in mg/kg) are 0.5 for cypermethrin, 0.1 for deltamethrin, 1.0 for fenvalerate and also 1.0 for permethrin.

Table 20. Residues of cyfluthrin in leeks (Internal Residue Reports, Bayer AG, Leverkusen, FRG)

Formulation	Days after last application	Country[1]	Residue levels (mg/kg) at harvest or at days after last application	Dosage a.i. (kg/ha)
50 EC	0/3/7/14	B	0.10/0.06/<0.04/<0.04	4 × 0.015
50 EC	0/1/3/7/14	NL	0.68/0.68/ 0.47/ 0.53/ 0.43	1 × 0.025
50 EC	10	F	<0.04	5 × 0.009
50 EC	10	F	0.12	5 × 0.009

[1] B = Belgium; NL = The Netherlands; F = France

Table 21a. Residues of pyrethroids in beans

Compound	Formulation	Days after last application	Country[1]	Residue levels (mg/kg) at harvest or at days after last application	Dosage a.i. (kg/ha)	Remarks (Ref.)
Cypermethrin	25 EC	0/3/7/10/15/21	IND	1.0/0.53/0.24/0.07/0.03/ND	2 × 0.074	100)
		0/3/7/10/15/21		1.9/0.78/0.50/0.13/0.04/ND	2 × 0.148	
Deltamethrin		0/3/7/10/15/21	IND	0.36/0.19/0.05/ND/ND/ND	2 × 0.011	100)
		0/3/7/10/15/21		0.72/0.36/0.12/0.07/ND/ND	2 × 0.022	
	^{14}C-Exp.	—	CAN	extractable ^{14}C, % bound ^{14}C, %		101)
				shoot $31.4^a/73.5^b$ $3.0^a/9.7^b$		
				root $0.6^b/$ 1.5^b $0.1^a/0.3^b$		
Fenvalerate	20 EC	0/3/7/10/15/21	IND	0.99/0.58/0.29/0.14/0.07/ND	2 × 0.074	100)
		0/3/7/10/15/21		2.23/0.90/0.60/0.20/0.09/ND	2 × 0.148	
	20 EC	1 hr/3/5/7/10/15	IND	0.71/0.45/0.36/0.26/0.06/ND	0.05	102)
		1 hr/3/5/7/10/15		1.21/0.67/0.54/0.36/0.11/0.02	0.10	
Permethrin	25 EC	0/3/7/10/15/21	IND	1.13/0.62/0.34/0.14/0.06/0.02	2 × 0.111	100)
		0/3/7/10/15/21		2.31/1.19/0.67/0.23/0.11/0.04	2 × 0.222	

[1] IND = India; CAN = Canada; ND = not detectable;
[a] ^{14}C-labeled at the methyl position; [b] ^{14}C-labeled at the benzylic position (further details see text).

Table 21b. Residues of cyfluthrin in snap beans (Table cont.) (Internal Residue Reports, Bayer AG, Leverkusen, FRG)

Formulation	Days after last application	Country[1]	Residue levels (mg/kg) at harvest or at days after last application	Dosage a.i. (kg/ha)
50 EC	3	F	<0.02	1 × 0.015
50 EC	3	F	<0.02	1 × 0.015

[1] F = France

The pre-harvest intervals are therefore between zero and 8 days depending upon the dosage of active ingredient applied [100].

The values reported under [101] should actually not be part of this compilation. Nevertheless they are listed because here the very interesting aspect of bound residues on bean plants is discussed. The percentage of bound residues in shoots is relatively low, namely between 3.0 (methyl label) and 9.7 % (benzylic label). It is interesting that after incubation with enzymes (beta-glucosidase, cellulase and protease) only portions of approx. 50 % (methyl label) or 30 % (benzylic label), resp., of the bound residues were released. The choice of the buffer system in this study was of great importance with the incubation of bound residues with enzymes because the release of active ingredient can be very different when using different buffer solutions.

When applying fenvalerate to pods of French beans, a safety interval of 7 days was observed [102]. The relatively high persistence of the pyrethroid can be at least partly attributed to the relatively high photostability of fenvalerate. The tolerance limit of the active ingredient on French beans is 0.2 mg/kg [102].

2.2.1.5.2 Peas, Lentils and Soybeans

Residue levels of pyrethroids on peas can be taken from Table 22. The respective values for cypermethrin, deltamethrin, fenvalerate and permethrin are summarized under [103]. There was no translocation of any of the 4 insecticides into green peas. The MRLs for grains are not exceeded in any case. The reduction of the initial residues was the slowest in case of deltamethrin and the quickest in case of cypermethrin. Seven days after application the residues in pea pods were below the MRL value. The residues in chick-peas (Cicer arietinum L.) were determined only at the time of harvest.

Table 22a. Residues of cyfluthrin in peas (Internal Residue Reports, Bayer AG, Leverkusen, FRG)

Formulation	Days after last application	Country[1]	Residue levels (mg/kg) at harvest or at days after last application	Dosage a.i. (kg/ha)
50 EC	99	F	<0.01	0.015, 2 applications
50 EC	99	F	<0.01	0.015, 2 applications

[1] F = France

Table 22b. Residues of pyrethroids in peas

Compound	Formulation	Days after last application	Country[1]	Residue levels (mg/kg) at harvest or at days after last application	Dosage a.i. (kg/ha)	Remarks (Ref.)
Cypermethrin		0/2/5/7 0/2/5/7 40	UK	pod: 0.18/0.14/0.06/0.03 pea: —/ND/ND/ND grain: ND; straw: 0.22	0.0375	[103]
				chick-pea: 0.032 (grain)/2.09 (straw) green gram: 0.02 (grain)/2.02 (straw) black gram: 0.02 (grain)/0.93 (straw)	0.0075%	
	25 EC	1 hr/3/7/15/grain at harvest	IND	green gram: 0.62/0.25/0.17/0.05/0.06	0.05	[104]
Deltamethrin	n.g.	0/2/5/7 0/2/5/7 40	UK	pod: 0.17/0.14/0.13/0.06 pea: —/ND/ND/ND grain: ND; straw: 0.08	0.025	[103]
				chick-pea: ND (grain)/1.42 (straw) green gram: ND (grain)/0.56 (straw) black gram: ND (grain)/0.23 (straw)	0.075%	
Fenpropanate	10 EC	1 hr/3/7/15/grain at harvest	IND	green gram: 0.69/0.45/0.22/0.14/0.11 cowpea: 0.39/0.25/0.12/0.11/0.03	0.05	[104]
Fenvalerate	n.g.	0/2/5/7 0/2/5/7 40	UK	pod: 0.72/0.62/0.23/0.13 pea: — ND/ND/ND grain: ND; straw: 0.37	0.10	[103]
				chick-pea: 0.01 (grain)/1.0 (straw) green gram: 0.02 (grain)/1.4 (straw) black gram: 0.02 (grain)/0.85 (straw)	0.02%	
	20 EC	1 hr/3/7/15/grain at harvest	IND	green gram: 0.75/0.42/0.25/0.12/0.08 cowpea: 0.38/0.23/0.09/0.04/0.02	0.05	[104]

Fenvalerate	n.g.	0/3/5/10/15	IND	2.8/1.7/1.0/0.80/0.30, 2nd spray 2.6/1.3/0.90/0.60/0.20/grain at harvest: ND	2 × 0.02%	105)
	n.g.	0/3/7/10/15/ straw at harvest: dry pod cover: grain:		1.54/0.86/0.76/0.51/0.21[a] 0.10 0.05 0.02	2 × 0.14	106)
Flucythrinate	10 EC	1 hr/3/7/15/grain at harvest	IND	green gram: 0.74/0.51/0.27/0.09/0.12 cowpea: 0.49/0.31/0.19/0.11/0.06	0.05	104)
Fluvalinate	25 EC	1 hr/3/7/15/grain at harvest	IND	green gram: 0.65/0.46/0.21/0.17/0.11 cowpea: 0.30/0.20/0.07/0.04/0.02	0.05	104)
Permethrin	n.g.	0/2/5/7 0/2/5/7 40	UK	pod: 0.56/0.50/0.19/0.11 pea: —/ND/ND/ND* grain: ND; straw: 0.09	0.075	103)
				chick-pea: 0.01 (grain)/2.1 (straw) green gram: 0.02 (grain)/1.1 (straw) black gram: 0.01 (grain)/0.75 (straw)	0.015%	
	EC	0/1/3/7	USA	0.98/0.69/0.59/0.18	0.112	18)
	n.g.	0/5/10/15 straw: dry pod cover: grain:	IND	2.43/1.72/1.39/0.91* 1.24 1.23 0.46	2 × 0.14	107)

[1] IND = India; UK = United Kingdom; n.g. = not given; ND = <0.001 mg/kg;
[a] all the values given are mean values of three replicates

Table 23. Residues of bioethanomethrin in lentils

Compound	Formulation	Days after last application	Country	Residue levels (mg/kg) at harvest or at days after last application	Dosage a.i. (kg/ha)	Remarks (Ref.)
Bioethanometrin	EC	0/66/168, 2nd appl., 0/72	USA	0.40/0.05/ND 0.67/ND	2 × 0.7	[18]
		at harvest		no residues found in lentils		

Residues on lentil foliage; the values given are mean values calculated via formation of the trichloroacetate ester- and trichloroethanol ester-derivatives; ND = not detectable

In that case relatively high residues were found in the straw. Translocation of the pyrethroids into the grain was observed — with the exception of deltamethrin.

A similar situation is known from green gram (Vigna radiata L.) and black gram (Vigna mungo L.). The residues appearing in the grains remain within safe limits [103].

Residues of cypermethrin, fenpropanate, fenvalerate, flucythrinate and fluvalinate in cowpeas (Vigna unguiculata L.) and in green gram (Vigna radiata L.) persisted for < 15 days on both crops and translocated to the grain at the time of harvest [104].

Dissipation order: cypermethrin > fenvalerate > fenpropanate > flucythrinate > fluvalinate.

The initial deposits were < 2 mg/kg (which corresponds to the MRL value).

Under [105] fenvalerate residues in red gram (Cajanus cajan L.) are listed in Table 22. The two applications were made at the time of pod formation at an interval of 15 days.

According to [106] fenvalerate residues were determined in chickpeas (Cicer arietinum L.) Two applications were made: the first at the flowering stage and the second 20 days later. If the chickpea pods were washed under running water for one minute, the residues declined from an initial level of 1.54 mg/kg to 0.96 mg/kg. After cooking only 0.71 mg/kg fenvalerate residues were detectable. At the time of harvest the residues were < 1 mg/kg.

The values given under [107] reflect residues of permethrin on chickpeas (Cicer arietinum L.). The respective applications were made at the flowering stage and 20 days later.

The extrapolated pre-harvest interval of 50 days for permethrin on peas results from the low MRL of 0.1 mg/kg [107].

Further permethrin residues on green peas are reported under [18] in Table 22. In this test the residues declined to approx. 18 % of the initial values within 7 days after treatment.

Some residue data of bioethanomethrin in lentils and cyfluthrin in soybeans are presented in Tables 23 and 24, respectively.

2.2.1.6 Root Vegetables

2.2.1.6.1 Celery/Celeriac

The residue situation of pyrethroids in celery is summarized in Table 25.

According to [97] cypermethrin, fenvalerate and permethrin were applied 3 times at weekly intervals at a rate of 0.08 kg a.i./ha each. Negligible levels of the 3 substances (< 0.1 mg/kg) were attained within a safety interval of 8–14 days after the last application [97]. According to [121] fenvalerate was not degraded within 14 days after treat-

Table 24. Residues of cyfluthrin in soybeans (Internal Residue Report, Bayer AG, Leverkusen, FRG)

Formulation	Days after last application	Country[1]	Residue levels (mg/kg) at harvest or at days after last application	Dosage a.i. (kg/ha)
50 SL	15/35	BR	< 0.01/ < 0.01	0.02 seeds

[1] BR = Brazil

Table 25. Residues of cypermethrin, fenvalerate and permethrin in celery

Compound	Formulation	Days after last application	Country[1]	Residue levels (mg/kg) at harvest or at days after last application	Dosage a.i. (kg/ha)	Remarks (Ref.)
Cypermethrin	Ripcord, 400 g/l	0/1/3/7/10/14	CAN	0.27/0.21/0.23/0.06/0.07/0.10	3 × 0.08	[97]
Fenvalerate	Belmark, 300 g/l	0/1/3/7/10/14	CAN	0.35/0.30/0.25/0.20/0.30/0.10	3 × 0.08	[97]
Permethrin	Ambush, 500 g/l	0/1/3/7/10/14	CAN	0.29/0.20/0.16/0.15/0.12/0.04	3 × 0.08	[97]
Fenvalerate	n.g.	0/1/3/7/10/14	CAN	0.12/0.14/0.25/0.14/0.19/0.17	3 × 0.08	[121].

[1] CAN = Canada; n.g. = not given

Table 26. Residues of permethrin in carrots

Compound	Formulation	Days after last application	Country[1]	Residue levels (mg/kg) at harvest or at days after last application	Dosage a.i. (kg/ha)	Remarks (Ref.)
Permethrin	Granular	n.g.	CAN	<0.005	0.56 and 1.12	[94]

= Canada; n.g. = not given

ment. Based on these varying findings the residue situation of this substance on celery does not seem to be completely clarified.

2.2.1.6.2 Carrots, Potatoes and Sugar Beets

After a soil application of a granular formulation of permethrin, residues of the substance did not occur in the edible portions of carrots (Table 26). This means that a translocation of the active ingredient from the soil into the root vegetable did not occur [94].

Table 27. Residues of cyfluthrin in potatoes (Internal Residue Reports, Bayer AG, Leverkusen, FRG)

Formulation	Days after last application	Country[1]	Residue levels (mg/kg) at harvest or at days after last application	Dosage a.i. (kg/ha)
50 EC	54	P	<0.01	2 × 0.03
25 EC	15	E	<0.02	1 × 0.025
50 EC	74	D	<0.01	1 × 0.025
50 EC	63	D	<0.01	1 × 0.025
50 EC	35	D	<0.01	1 × 0.025
50 EC	35	D	<0.01	1 × 0.025

[1] P = Portugal; E = Spain; D = Federal Republik of Germany

Table 28. Residues of cyfluthrin in sugar-beets including leaves and processed products (Internal Residue Reports, Bayer AG, Leverkusen, FRG)

Formulation	Days after last application	Country[1]	Residue levels (mg/kg) at harvest or at days after last application	Dosage a.i. (kg/ha)
25 SL	152	F	leaves 0.02; beets 0.02	1 × 0.025
25 SL	105	F	leaves 0.02; beets 0.02	1 × 0.025
50 EC	3	GB	beets: <0.02	1 × 0.025
240 EC	0	USA	beets: 0.06	6 × 0.208
240 EC	0	USA	pressed pulp: <0.04	6 × 0.208
240 EC	0	USA	2: <0.04	6 × 0.208
240 EC	0	USA	thick juice: <0.04	6 × 0.028
240 EC	0	USA	thin juice: <0.02	6 × 0.208
240 EC	0	USA	molasses: <0.02	6 × 0.208
240 EC	0	USA	A-sugar: <0.04	6 × 0.208
240 EC	0	USA	refined sugar: <0.04	6 × 0.208
240 EC	0	USA	beets: 0.04	6 × 0.208
240 EC	0	USA	pressed pulp: <0.04	6 × 0.208
240 EC	0	USA	2: <0.04	6 × 0.208
240 EC	0	USA	thick juice: <0.04	6 × 0.208
240 EC	0	USA	thin juice: <0.02	6 × 0.208
240 EC	0	USA	molasses: <0.02	6 × 0.208
240 EC	0	USA	A-sugar: <0.04	6 × 0.208
240 EC	0	USA	refined sugar: <0.04	6 × 0.208

[1] F = France; GB = United Kingdom; [2] sludge from liming and carbonation

Table 29. Residues of permethrin in turnip roots and turnip greens

Compound	Formulation	Days after last application	Country	Residue levels (mg/kg) at harvest or at days after last application	Dosage a.i. (kg/ha)	Remarks (Ref.)
Permethrin	EC	1/3/5	USA	0.07/0.02/0.03 (roots)[a]	8 × 0.112	82)
		1/3/5		0.05/0.05/0.05 (roots)[a]	8 × 0.224	
		1/3/5		6.33/1.43/1.66 (greens)	8 × 0.112	
		1/3/5		6.69/1.43/1.32 (greens)	8 × 0.224	

[a] Mean values from two different locations

Of interest is the finding also reported under [94] that permethrin is obviously not phytotoxic to carrots while the yields of onions are strongly decreased: the untreated control yielded 22 tons of onions/ha while the treated area yielded only approximately 12–13 tons/ha.

Residue data of permethrin in carrots and of cyfluthrin in potatoes and sugar beets are summarised in Tables 26–28 with Table 28 showing detailed data on residues after processing.

2.2.1.6.3 Turnips

Residues of permethrin in turnip roots and turnip greens are listed in Table 29 [82]. The data were obtained after 8 applications at weekly intervals.

In the turnip greens the residues declined by approx. 80% within 3 days. Only small amounts of permethrin were translocated to the roots. The average values in the roots were <0.1 mg/kg. This residue situation in turnip roots is therefore similar to that reported under 2.2.1.4.1 and 2.2.1.6.2 under Ref. [94] for onions or carrots respectively.

2.2.1.6.4 Red Beets

The values reported in Table 30 under [98] are based on a study concerning the uptake of fenvalerate by red beets (table beets) after soil treatment. The results show that red beets take up only a very small amount of the substance from the soil. The amount of active ingredient applied in this test was considerable: 2.2 kg a.i./ha!

Regrettably the data on "vegetables" presented in Table 31 cannot be assigned to any specific species of vegetable. They are listed just for the sake of completeness.

2.2.2 Fruits

2.2.2.1 Berries

2.2.2.1.1 Blueberries

Residue data of cypermethrin and permethrin in blueberries are reported in Table 32 under [108] and [61]. Concerning the values listed under [108] it should be mentioned that the residue situation is also influenced by the time-span between fruit set and harvest. Moreover, the size of the berries is obviously another factor influencing the levels of the residue.

The applications were made at weekly intervals. As a result of these studies it can be said that cypermethrin and permethrin produce persistent residues on blueberries.

Table 30. Residues of fenvalerate in red beets

Compound	Formulation	Days after last application	Country	Residue levels (mg/kg) at harvest or at days after last application	Dosage a.i. (kg/ha)	Remarks (Ref.)
Fenvalerate	n.g.	30/120/1 year[a]	USA	0.14/0.02/0.03 (leaf) 0.29/0.04/0.04 (root)	2.2	[98)]

n.g. = not given;
[a] soil treatment; planting 30 days, 120 days and 1 year after treatment (see text)

Table 31. Residues of pyrethroids in/on "vegetables"[a] from Ontario region, Canada

Compound	Formulation	Days after last application	Country[1]	Residue levels (mg/kg) at harvest or at days after last application	Dosage a.i. (kg/ha)	Remarks (Ref.)
Cypermethrin Deltamethrin Fenvalerate Permethrin	unknown[b]	unknown	CAN	<0.005 <0.005 <0.005 <0.005	unknown	[122)]

[1] CAN = Canada;
[a] kind not specified; [b] Official market control for pestic. res. on Ontario-grown vegetables in 1984 and 1985. 23 vegetable samples were analyzed for the synthetic pyrethroids. Commodities tested: asparagus, beans, carrots, caulifower, cucumbers, onions, potatoes, sweet corn, and tomatoes

Table 32. Residues of pyrethroids in blueberries

Compound	Formulation	Days after last application	Country[1]	Residue levels (mg/kg) at harvest or at days after last application	Dosage a.i. (kg/ha)	Remarks (Ref.)
Cypermethrin	10 EC	at different stages of harvest	NZ	0.15–0.69[a]	3 × 0.0075	108)
Permethrin	50 EC	at different stages of harvest	NZ	0.87–2.56[a]	3 × 0.020	108)
	25 WP	at harvest	CAN	<0.05	0.40	61)

[1] NZ = New Zealand; CAN = Canada.
[a] the residues depend on the percentage fruit-set at the time of spraying (for further details see text).

Table 33. Residues of permethrin in raspberries

Compound	Formulation	Days after last application	Country	Residue levels (mg/kg) at harvest or at days after last application	Dosage a.i. (kg/ha)	Remarks (Ref.)
Permethrin	EC	1/3/5/7 1/3/5/7	USA	0.30/0.29/0.26/0.13 0.70/0.66/0.33/0.20	2 × 0.224 2 × 0.448	82)

The findings reported under [61] concerning residues of permethrin on blueberries do not agree with the above results. No residues were detectable in the berries (<0.05 mg/kg). In this test, however, the blueberry plants were treated when the shoots were up to 3 cm high.

Of interest is the finding that according to [61] an increased crop yield resulted after a permethrin treatment.

2.2.2.1.2 Raspberries

According to Table 33 [82] the permethrin residues decline within 7 days after the last application by approx. 60–70% in comparison to the initial residues. The applications for this trial were made at 2-week-intervals. Parallel to analyses for residues of unchanged parent compound, the samples from this trial were additionally analysed for residues of the two metabolites dichlorovinyl acid and metaphenoxybenzyl alcohol. No residues were detectable in this case (<0.05 mg/kg).

2.2.2.1.3 Strawberries

According to Table 34 and Ref. [121] the MRL of 0.1 mg/kg of pyrethroid residues on strawberries is exceeded only in the case of cypermethrin on the 7th day after the last application at a rate of 0.14 kg a.i./ha.

2.2.2.2 Pome Fruits

2.2.2.2.1 Apples

The following remarks are to be made concerning the residue data of pyrethroids on apples (Table 35).

The values reported under [109] originate from the UK market control in 1981. In this case 7 samples of apples were analysed for residues of cypermethrin, deltamethrin, fenvalerate and permethrin. Only cypermethrin and fenvalerate could be detected in one sample each.

In 1983 42 apple samples were analysed for pyrethroid residues in the UK in the course of the official market control [111]. Only in one sample a positive finding was reported: 0.034 mg/kg fenpropathrin. Also this value was markedly below the "reporting limit" of 0.10 mg/kg for pyrethroids.

According to [110] the residues of the 4 pyrethroids cypermethrin, deltamethrin, fenvalerate and permethrin in whole apples are negligibly low. Residues found on peels are markedly higher than those found in the pulp. According to [83] fenvalerate was applied at 20 day intervals.

The fenvalerate residues reported in the Table under [53] are based on a cooperative study concerning processed apple products and by-products. Five laboratories were involved in this study. The various laboratories concerned indicated different limits of determination for juice and sauce.

The values presented by the individual laboratories were well in agreement. The use of fenvalerate on apples which are provided for the production e.g. of baby food "should pose a minimal health hazard".

Table 34a. Residues of pyrethroids in strawberries

Compound	Formulation	Days after last application	Country[1]	Residue levels (mg/kg) at harvest or at days after last application	Dosage a.i. (kg/ha)	Remarks (Ref.)
Cyfluthrin	n.g.	7/10/14/21	CAN	0.04/0.03/0.03/0.01[a]	3 × 0.015	121)
Cypermethrin	n.g.	7/10/14/21	CAN	0.05/0.03/0.07/0.007[b]	3 × 0.07	121)
	n.g.	7/10/14/21	CAN	0.12/0.09/0.07/0.02[c]	3 × 0.14	121)
Deltamethrin	n.g.	7/10/14/21	CAN	0.01/0.003/0.002/0.001	3 × 0.01	121)
	n.g.	7/10/14/21	CAN	0.009/0.008/0.006/ND	3 × 0.015	121)
Fenvalerate	n.g.	7/10/14/21	CAN	0.07/0.27/0.03/0.015[d]	3 × 0.10	121)
Permethrin	n.g.	7/10/14/21	CAN	0.08/0.08/0.12/0.02[c]	3 × 0.14	121)

[1] CAN = Canada; ND = not detectable; [a] Mean values from 3 replicates in three different years; [b] mean values from 4 replicate experiments; [c] mean values from 2 replicates in two different years; [d] mean values from two replicate experiments

Table 34b. Residues of cyfluthrin in strawberries (Table cont.) (Internal Residue Reports, Bayer AG, Leverkusen, FRG)

Formulation	Days after last application	Country[1]	Residue levels (mg/kg) at harvest or at days after last application	Dosage a.i. (kg/ha)
50 EC	3	E	0.04	1 × 0.025
50 EC	3	E	0.02	1 × 0.025

[1] E = Spain

References [112-114] refer to a study concerning the influence of insecticides upon fruit quality characteristics in apples (Golden Delicious). In the apples harvested 1980 and 1981 no residues were detectable while in the apples harvested in 1982 minimal residues were found which were, however, all below 0.01 mg/kg. The occurrence of these residues in the 1982 crop, though they were very low, can be explained by the fact that this crop was harvested after 7 permethrin treatments while 3–4 treatments per year are common in the area of North Germany. The residue situation of cis- and trans permethrin was identical.

2.2.2.2.2 Pears

The residue situation of pyrethroids on pears can be taken from Table 36.

According to [48] permethrin residues in pears occur only in the peel. The residues in the entire fruit are below the FAO/WHO MRL for pome fruits (2 mg/kg). The values reported under [48] were determined by means of high pressure liquid chromatography (HPLC). The transisomer partly constituted a markedly greater proportion of the total residues (cis/trans) than the cis-isomer. [109] and [111] show the results of the official market control analyses in the United Kingdom in 1981 and 1983. According to [109] 5 samples each were analysed for residues of the 4 pyrethroids cypermethrin, deltamethrin, fenvalerate and permethrin. Only in one sample each small amounts of residues of deltamethrin and fenvalerate resp. were detected in this case.

The study according to [111] is based on 37 pear samples. Only in two of these samples small amounts of permethrin residues of 0.042 mg/kg each were found.

2.2.2.3 Stone Fruits

2.2.2.3.1 Cherries, Peaches, Plums/Damson and Nectarines

Table 37 shows the residue situation of cyfluthrin in sour cherries.

Residue results of pyrethroids in peaches are summarized in Table 38. The results of the official market control in the United Kingdom in 1983 are presented under [111]. Three samples were analysed for fenvalerate residues with negative results (< 0.024 mg/kg).

The results presented under [115] are based on a study to determine the dissipation of cis/trans-permethrin on peach leaves. The individual results can be taken from Fig. 7. Twenty-one after the first application (June 2, 1976) still 10% of the intial deposits were present on the leaves, the comparable value for the second application (July 14, 1976) is 41%.

During the period of 21 days the two isomers were degraded at about the same rate. According to these results the half-life of permethrin on peach leaves can be considered to be 7–10 days.

Data on residues of cyfluthrin in plums/damson plums and nectarines are presented in Tables 39–41.

2.2.2.4 Grapes

Residues of pyrethroids in grapes are reported in Table 42. The results show that most of the pyrethroids listed persist on grapes for more than 3 weeks. Based on tolerance limits of 0.2 mg/kg for fenvalerate and 0.5 mg/kg for the 3 other pyrethroids listed, the resulting safety intervals [116] are 9.5 days for fenvalerate, 2.6 days for permethrin and

Table 35a. Residues of pyrethroids in apples

Compound	Formulation	Days after last application	Country[1]	Residue levels (mg/kg) at harvest or at days after last application	Dosage a.i. (kg/ha)	Remarks (Ref.)
Cypermethrin	unknown[a]	unknown	UK	0.005	unknown	109)
Deltamethrin	unknown[a]	unknown	UK	"no residues"	unknown	109)
Fenvalerate	unknown[a]	unknown	UK	0.003	unknown	109)
Permethrin	unknown[a]	unknown	UK	"no residues"	unknown	109)
Cypermethrin	40 EC	35	CAN	peel: 1.94/pulp: 0.10 whole apple: 0.19	3 × 0.05	110)
	12.5 WP	43	CAN	peel: 0.91/pulp: 0.002 whole apple: 0.004	2 × 0.10	110)
	20 WP	43	CAN	peel: 0.78/pulp: 0.002 whole apple: 0.005	2 × 0.05	110)
Deltamethrin	25 EC	43	CAN	peel: 0.19/pulp: 0.001 whole apple: 0.001	2 × 0.0115	110)
Fenvalerate	30 EC	35	CAN	peel: 1.45/pulp: 0.05 whole apple: 0.12	2 × 0.07	310)
	30 EC	43	CAN	peel: 1.62/pulp: 0.004 whole apple: 0.082	2 × 0.07	110)
Permethrin	25 WP	35	CAN	peel: 2.14/pulp: 0.02 whole apple: 0.12	3 × 0.200	110)
	25 WP	43	CAN	peel: 0.68/pulp: 0.003 whole apple: 0.004	2 × 0.150	110)
Fenpropathrin	unknown[a]	unknown	UK	0.034	unknown	111)
Fenvalerate	"20%"-formul.	0/8/30	ISR	1.1/0.6/0.45	5 × 0.45	83)
	EC (1.09 kg a.i./gal)	processing study	USA	whole apples: 0.4 sauce: <0.25–<0.003[b] juice: <0.25–<0.005[b] pomace: 2 peels plus cores: 1.5	9 foliar applicat. (60 g or 120 g/100 gal	53)
Permethrin	EC	28 (at harvest)	D	<0.01	"7 applic."	112/113/114)

[1] CAN = Canada; UK = United Kingdom; ISR = Israel; D = Federal Republic of Germany;
[a] official market control in UK 1981 109) and 1983 111), for further details see text; [b] cooperative study (see text)

Table 35b. Residues of cyfluthrin in apples (Table con.) (Internal Residue Reports, Bayer AG, Leverkusen, FRG)

Formulation	Days after last application	Country[1]	Residue levels (mg/kg) at harvest or at days after last application	Dosage a.i. (kg/ha)
50 EC	0/4/7/14/21	D	0.10/0.09/0.08/0.06/0.04	5 × 0.375
50 EC	0/4/7/14/21	D	0.09/0.05/0.04/0.03/0.03	5 × 0.375
50 EC	0/4/7/14/21	D	0.06/0.02/0.05/0.04/0.03	4 × 0.375
50 SL	0/4/7/14/21	I	0.29/0.25/0.21/0.18/0.19	5 × 0.05
50 SL	0/4/7/14/21	D	0.12/0.09/0.08/0.08/0.07	5 × 0.0375
50 SL	0/4/7/14/21	I	0.22/0.20/0.13/0.09/0.09	4 × 0.05
50 SL	0/4/7/14/21	D	0.10/0.08/0.08/0.03/0.02	4 × 0.0375
50 SL	0/4/7/14/21	I	0.17/0.14/0.18/0.10/0.15	5 × 0.05
50 EC	0/3/5/7/10/14	D	0.11/0.06/0.05/ 0.09/ 0.08/<0.01	5 × 0.023
50 EC	0/3/5/7/10/14	D	0.04/0.01/0.06/<0.01/<0.01/<0.01	5 × 0.023
50 EC	0/3/5/7/10/14	D	0.04/0.05/0.05/ 0.05/ 0.05/ 0.03	5 × 0.015
50 EC	0/3/5/7/10/14	D	0.07/0.14/0.05/ 0.04/ 0.04/ 0.02	5 × 0.023
50 EC	0/7/10/14/21	F	0.13/0.08/0.06/ 0.08/ 0.07	10 × 0.05
12.5 EC	0/3/5/7/10/14	D	0.16/0.14/0.10/ 0.09/ 0.10/ 0.08	5 × 0.038
12.5 EC	0/3/5/7/10/14	D	0.11/0.12/0.11/ 0.12/ 0.14/ 0.08	5 × 0.038
12.5 EC	0/3/5/7/10/14	D	0.06/0.05/0.05/ 0.04/ 0.04/ 0.06	5 × 0.038
12.5 EC	0/3/5/7/10/14	D	0.10/0.05/0.09/ 0.07/ 0.06/ 0.06	5 × 0.038
50 EC	10	F	0.06	4 × 0.024
12.5 EC	0/3/5/7/10/14	D	0.10/0.08/0.08/ 0.06/ 0.06/ 0.06	5 × 0.038
12.5 EC	0/3/5/7/10/14	D	0.10/0.10/0.10/ 0.10/ 0.10/ 0.06	5 × 0.038
12.5 EW	0/3/5/7/10/14	D	0.04/0.08/0.05/0.04/0.06/0.04	5 × 0.038
12.5 EW	0/3/5/7/10/14	D	0.10/0.05/0.06/0.05/0.05/0.04	5 × 0.038
	60		0.05	
	0/7		0.073/0.052	
	0/7		0.13/0.08	

[1] D = Federal Republic of Germany; I = Italy; F = France

Table 36a. Residues of pyrethroids in pears

Compound	Formulation	Days after last application	Country[1]	Residue levels (mg/kg) at harvest or at days after last application	Dosage a.i. (kg/ha)	Remarks (Ref.)
Cypermethrin	unknown[a]	unknown	UK	no residues found	unknown	109)
Deltamethrin	unknown[a]	unknown	UK	0.005	unknown	109)
Fenvalerate	unknown[a]	unknown	UK	0.041	unknown	109)
Permethrin	unknown[a]	unknown	UK	no residues found	unknown	109)
Permethrin	n.g.	at harvest	UK	peel: 1.6 (cis- + trans-isomers) peel: 0.1 (cis- + trans-isomers) peel: 7.1 (cis- + trans-isomers)	n.g.	48)
	unknown[b]	unknown	UK	0.042/0.042	unknown	111)

[1] UK = United Kingdom; n.g. = not given;
[a] results of market control in UK in 1981;
[b] results of market control in UK in 1983; for further details concerning [a] and [b] see text

Table 36b. Residues of cyfluthrin in pears (Table cont.) (Internal Residue Reports, Bayer AG, Leverkusen, FRG)

Formulation	Days after last application	Country[1]	Residue levels (mg/kg) at harvest or at days after last application	Dosage a.i. (kg/ha)
50 SL	0/7/10/14/21	F	0.04/0.04/0.05/0.04/0.03	1 × 0.03
12.5 EC	0/3/5/7/10/14	D	0.09/0.09/0.09/0.08/0.08/0.08	5 × 0.038
12.5 EC	0/3/5/7/10/14	D	<0.01/0.09/0.08/0.09/0.09/0.10	5 × 0.038
12.5 EC	0/3/5/7/10/14	D	0.08/0.05/0.06/0.05/0.05/0.03	5 × 0.025
12.5 EC	0/3/5/7/10/14	D	0.08/0.04/0.04/0.04/0.03/0.02	5 × 0.025
12.5 EC	0/3/5/7/10/14	D	0.08/0.07/0.08/0.06/0.06/0.03	5 × 0.038
12.5 EC	0/3/5/7/10/14	D	0.10/0.11/0.09/0.11/0.09/0.10	5 × 0.038
50 EC	15	F	0.05	3 × 0.035
50 EC	21	I	0.09	
50 EC	21	I	0.07	
50 EC	21	I	0.07	
12.5 EW	0/3/5/7/10/14	D	0.06/0.05/0.03/0.03/0.03/0.02	5 × 0.038
12.5 EW	0/3/5/7/10/14	D	0.13/0.09/0.08/0.08/0.07/0.06	5 × 0.038

[1] F = France; D = Federal Republic of Germany; I = Italy

Table 37. Residues of cyfluthrin in sour cherriers (Internal Residue Reports, Bayer AG, Leverkusen, FRG)

Formulation	Days after last application	Country[1]	Residue levels (mg/kg) at harvest or at days after last application	Dosage a.i. (kg/ha)
50 EC	0/3/5/7/10/14	D	0.11/0.07/0.07/0.05/0.04/0.02	3 × 0.015
50 EC	0/3/5/7/10/14	D	0.22/0.13/0.15/0.10/0.06/0.04	3 × 0.015
50 EC	0/3/5/7/10/	D	0.22/0.12/0.12/0.10/0.09	3 × 0.015
50 EC	0/3/5/7/10/	D	0.19/0.20/0.14/0.14/0.14	3 × 0.015
50 EC	0/3/5/7/10	D	0.06/0.06/0.05/0.05/0.05	3 × 0.015
50 EC	0/3/5/7/10	D	0.10/0.08/0.05/0.06/0.03	3 × 0.015
50 EC	0/3/5/7/10	D	0.12/0.09/0.07/0.05/0.04	3 × 0.015
50 EC	0/3/5/7/10	D	0.06/0.05/0.04/0.02/0.02	3 × 0.015

[1] D = Federal Republic of Germany

0 days for cypermethrin and deltamethrin. The two last named active ingredients dissipate faster [116]

According to [83] the fenvalerate residues in grapes decline from 0.08 mg/kg to 0.025 mg/kg within 20 days after the last application (dosage 0.0375 kg/ha). When applying twice the rate (0.075 kg a.i./ha), the resulting residues were between 0.20 and 0.33 mg/kg, which did not decline during the period of observation. This discrepancy should be further investigated.

2.2.2.5 Oranges

Residues of cypermethrin, deltamethrin, fenvalerate and permethrin are presented in Table 43.

Table 38a. Residues of pyrethroids in peaches

Compound	Formulation	Days after last application	Country[1]	Residue levels (mg/kg) at harvest or at days after last application	Dosage a.i. (kg/ha)	Remarks (Ref.)
Fenvalerate	unknown[a]	unknown	UK	<0.024	unknown	[111]
Permethrin	25 WP	0/21	CAN	on peach leaves: 12.8/1.4-[b] after the first application	2 × 0.21	[115]
		0/21		on peach leaves: 16.6/5.4-[b] after the second application		

[1] UK = United Kingdom, CAN = Canada;
[a] results from the official market control in UK, 1983; [b] combined residues of cis- and trans-isomers (see text)

Table 38b. Residues of cyfluthrin in peaches (Table cont.) (Internal Residue Reports, Bayer AG, Leverkusen, FRG)

Formulation	Days after last application	Country[1]	Residue levels (mg/kg) at harvest or at days after last application	Dosage a.i. (kg/ha)
50 EC	15	F	0.04	3 × 0.02
50 EC	28	F	0.02	3 × 0.049
50 EC	11	F	0.14	3 × 0.042

[1] F = France

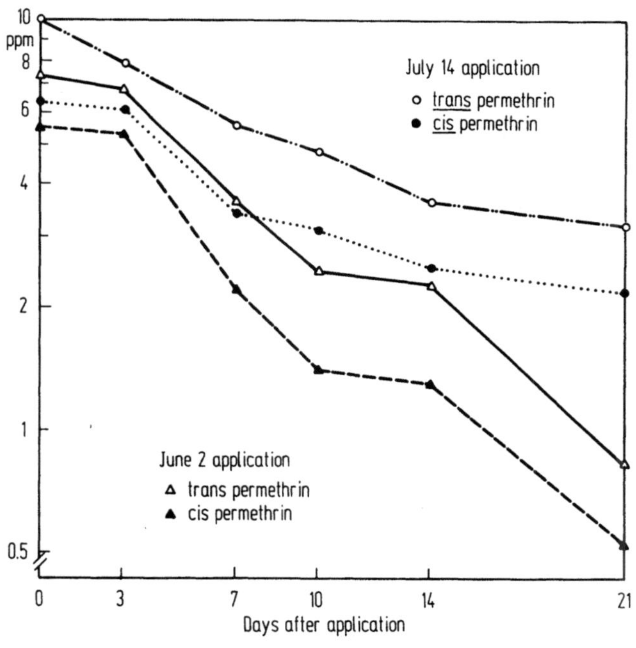

Fig. 7. Initial deposits and subsequent residues of cis and trans permethrin on peach leaves after the June 2 and July 14 spray applications (according to M. Chiba [115])

In case of the data reported under [117] the active ingredients were applied to run-off at the concentration mentioned using 2 litres of spray fluid per tree. The residues listed refer to the peel and are given in percent of the initial residues. No residues were detectable in the pulp. The order of persistence in the peel is as follows: fenvalerate > permethrin > cypermethrin > deltamethrin.

2.2.3 Cereals and Grain Products

2.2.3.1 Barley and Oats

In cereals pyrethroids are mostly used for stored product protection. Respective residues on barley can be taken from Table 44. The residue data for bioresmethrin and

Table 39. Residues of cyfluthrin in plums (Internal Residue Reports, Bayer AG, Leverkusen, FRG)

Formulation	Days after last application	Country[1]	Residue levels (mg/kg) at harvest or at days after last application	Dosage a.i. (kg/ha)
50 EC	0/3/5/7/10/14	D	0.13/0.11/0.06/ 0.05/ 0.05/ 0.05	5 × 0.023
50 EC	0/3/5/7/10/14	D	0.10/0.08/0.02/<0.02/<0.02/<0.02	5 × 0.023
12.5 EC	0/3/5/7/10/14	D	0.26/0.54/0.39/ 0.31/ 0.19/ 0.26	5 × 0.038
12.5 EC	0/3/5/7/10/14	D	0.02/0.04/0.04/ 0.02/ 0.03/ 0.01	5 × 0.038
50 EC	0/3/5/7/10	D	0.01/0.01/<0.01/<0.01/<0.01	5 × 0.023
50 EC	0/3/5/7/10	D	0.07/0.07/0.11/ 0.02/ 0.02	5 × 0.023
12.5 EW	0/3/5/7/10/14	D	0.24/0.16/0.20/ 0.07/ 0.06/ 0.03	5 × 0.038
12.5 EW	0/3/5/7/10/14	D	0.05/0.05/0.03/ 0.02/ 0.01/ 0.03	5 × 0.038

[1] D = Federal Republic of Germany

Table 40. Residues of cyfluthrin in damson plums (Internal Residue Reports, Bayer AG, Leverkusen, FRG)

Formulation	Days after last application	Country[1]	Residue levels (mg/kg) at harvest or at days after last application	Dosage a.i. (kg/ha)
12.5 EC	0/3/5/7/10/14	D	0.04/0.04/0.02/0.04/0.02/0.03	5 × 0.038
12.5 EC	0/3/5/7/10/14	D	0.05/0.06/0.07/0.06/0.12/0.13	5 × 0.038
12.5 EC	0/3/5/7/10/14	D	0.03/0.03/0.03/0.02/0.02/0.02	2 × 0.025/ 3 × 0.038
12.5 EC	0/3/5/7/10/14	D	0.10/0.10/0.08/0.07/0.07/0.09	5 × 0.038
12.5 EC	0/3/5/7/10/14	D	0.10/0.19/0.19/0.08/0.15/0.11	5 × 0.038
12.5 EC	0/3/5/7/10/14	D	0.09/0.09/0.05/0.05/0.06/0.06	5 × 0.038
50 EC	0/3/5/7/10/14	D	0.07/0.06/0.04/0.05/0.05/0.04	5 × 0.023
50 EC	0/3/5/7/10/14	D	0.30/0.14/0.11/0.16/0.08/0.07	5 × 0.023
50 EC	0/3/5/7/10	D	0.10/0.12/0.19/0.10/0.04	5 × 0.023
50 EC	0/3/5/7/10	D	0.02/0.02/0.02/0.02/0.01	5 × 0.023
50 EC	0/3/5/7/10	D	0.07/0.07/0.11/0.02/0.02	5 × 0.023
12.5 EC	0/3/5/7/10/14	D	0.24/0.16/0.20/0.07/0.06/0.03	5 × 0.038
12.5 EC	0/3/5/7/10/14	D	0.04/0.04/0.04/0.06/0.02/0.03	5 × 0.038
12.5 EC	0/3/5/7/10/14	D	0.18/0.09/0.08/0.13/0.07/0.07	5 × 0.038

[1] D = Federal Republic of Germany

Table 41. Residues of cyfluthrin in nectarines (Internal Residue Report, Bayer AG, Leverkusen, FRG)

Formulation	Days after last application	Country[1]	Residue levels (mg/kg) at harvest or at days after last application	Dosage a.i. (kg/ha)
50 EC	14	F	0.10	3 × 0.02

[1] F = France

Table 42a. Residues of pyrethroids in grapes

Compound	Formulation	Days after last application	Country[1]	Residue levels (mg/kg) at harvest or at days after last application	Dosage a.i. (kg/ha)	Remarks (Ref.)
Cypermethrin	n.g.	0/3/7/10/15/21	IND	0.35/0.22/0.13/0.08/0.06/ND	conc. (%): 0.0075	116)
				0.84/0.55/0.32/0.17/0.09/0.07	conc. (%): 0.01	
Deltamethrin	n.g.	0/3/7/10/15/21		0.13/0.11/0.08/0.05/ND/ND	conc. (%): 0.0015	
				0.17/0.13/0.08/0.06/ND/ND	conc. (%): 0.002	
Fenvalerate	n.g.	0/3/7/10/15/21		0.78/0.57/0.31/0.17/0.08/0.04	conc. (%): 0.015	
				1.22/0.84/0.41/0.24/0.11/0.07	conc. (%): 0.02	
Permethrin	n.g.	0/3/7/10/15/21		0.77/0.47/0.26/0.13/0.05/0.02	conc. (%): 0.015	
				1.35/0.73/0.43/0.19/0.10/0.04	conc. (%): 0.02	
Fenvalerate	n.g.	0/6/13/20	ISR	0.08/0.08/0.02/0.025 0.24/0.20/0.33/0.28	0.0375 0.075	83)

[1] IND = India; ISR = Israel; n.g. = not given; ND = not detectable

Table 42b. Residues of cyfluthrin in grapes, including juice and wine (Table cont.) (Internal Residue Reports, Bayer AG, Leverkusen, FRG)

Formulation	Days after last application	Country[1]	Residue levels (mg/kg) at harvest or at days after last application	Dosage a.i. (kg/ha)
50 EC	0/21/28/35	D	0.34/0.11/0.11/0.23/juice: 35 days: 0.02 wine: 35 days: <0.01	0.025/0.037 0.05 l/ha
50 EC	0/21/28/35	D	0.34/0.19/0.20/0.25/juice: 35 days: 0.01 wine: 35 days: <0.01	0.025/0.037 0.05 l/ha
50 EC	0/21/28/35	D	0.20/0.14/0.10/0.08/juice: 35 days: 0.01 wine: 35 days: <0.04	0.025/0.037 0.05 l/ha
100 EC	7/14	T	0.32/0.38	15 × 0.075
100 EC	7/14	T	0.16/0.14	15 × 0.075
100 EC	7/14	T	0.28/0.23	15 × 0.075
100 EC	7/14	T	0.59/0.62 } washing the grapes did not result in lower residue levels	15 × 0.075
100 EC	7/14	T	0.35/0.45	15 × 0.075
100 EC	7/14	T	0.28/0.23	15 × 0.075
50 SL	29	F	0.04	
50 EC	77	F	<0.01	3 × 0.02
50 EC	7	F	0.02	3 × 0.02

[1] D = Federal Republic of Germany; F = France; T = Thailand

Table 43. Residues of pyrethroids in oranges

Compound	Formulation	Days after last application	Country[1]	Residue levels (in %) at harvest or at days after last application	Dosage a.i. (kg/ha)	Remarks (Ref.)
Cypermethrin	n.g.	3/7/10/15/21	IND	44.8/10.9/1.7/0.0/—[a]	conc. (%): 0.0075	117)
				45.0/15.1/3.4/0.0/—	conc. (%): 0.01	
Deltamethrin	n.g.	3/7/10/15/21		51.1/15.5/0.0/—/—	conc. (%): 0.0015	
				49.3/19.1/0.0/—/—	conc. (%) 0.002	
Fenvalerate	n.g.	3/7/10/15/21		43.6/21.1/5.3/1.4/0.0	conc. (%): 0.015	
				42.0/19.7/7.0/2.3/1.1	conc. (%): 0.02	
Permethrin	n.g.	3/7/10/15/21		56.8/31.1/13.2/0.8/0.0	conc. (%): 0.015	
				47.6/24.6/13.2/2.4/0.3	conc. (%): 0.02	

[1] IND = India; n.g. = not given;
[a] for further details see text

Table 44a. Residues of pyrethroids in barley (stored product protection and storage stability study)

Compound	Formulation	Days after last application	Country[1]	Residue levels (mg/kg) at harvest or at days after last application	Dosage a.i. (kg/ha)	Remarks (Ref.)
Bioresmethrin	n.g.	3 months/6 months	AUS	4.0/2.5[a]	7 mg/kg of grain	[118]
		3 months/6 months		1.5/0.25 each after malting[b]		
Phenothrin	n.g.	3 months/6 months	AUS	—/4.6[a]	8 mg/kg of grain	[118]
		3 months/6 months		—/0.55 after malting[b]		

[1] AUS = Australia; n.g. = not given
[a] values after storage and losses during processing (for further details see text);
[b] 3-months-value fo bioresmethrin after home malting; 6-months-values for both pyrethroids after pilot commercial malting

Table 44b. Residues of cyfluthrin in barley (Table cont.) (Internal Residue Reports, Bayer AG, Leverkusen, FRG)

Formulation	Days after last application	Country[1]	Residue levels (mg/kg) at harvest or at days after last application	Dosage a.i. (kg/ha)
50 EC	264	F	grain: <0.01/straw: <0.02	1 × 0.025
50 EC	262	F	grain: <0.01/straw: <0.02	1 × 0.025

[1] F = France

Table 45. Residues of cyfluthrin in oats (Internal Residue Reports, Bayer AG, Leverkusen, FRG)

Formulation	Days after last application	Country[1]	Residue levels (mg/kg) at harvest or at days after last application	Dosage a.i. (kg/ha)
50 EC	95	S	<0.01	0.025 grain
50 EC	93	S	<0.01	0.025 grain

[1] S = Sweden

d-phenothrin reported under [118] were determined after a storage period of three and six months respectively. In addition, the fate of residues during processing was determined. It was found that commercial malting of barley resulted in a considerable degradation of the pyrethroids. Controlled conditions prevailed during the study: storage temperature 25 °C, rel. humidity 65 %.

After storage for 6 months the barley was malted in a pilot process which simulated commercial conditions. The differences between the values obtained after home malting and those obtained after pilot commercial malting can be explained by more intensive washing, wetter germination conditions and greater degree of heating in kilning of the pilot malt.

Some residue data of cyfluthrin in oats were reported from Sweden (Table 45).

2.2.3.2 Rye

Also the residue data of cyfluthrin, cypermethrin, deltamethrin, fenvalerate and permethrin reported in Table 46 are based on the use of the pyrethroids in stored product protection and include the residue situation after processing of the rye. The residue levels of the pyrethroids remaind practically constant over an observation period of 10 months independent of the storage temperature and the initial concentration.

The acitve ingredients are distributed in the processing products as follows: approx. 50 % in the flour, approx. 7 % in the coarse flour and approx. 40 % in the bran. If milling was repeated after 7 mohths storage the average contents found in the flour were 20–25 % lower while the contents in the bran were correspondingly higher. This observation suggests that the pyrethroids are slowly accumulated in the marginal layers of the grain.

After baking, 30–50 % of the intial pyrethroid residues contained in the rye grains were lost.

The higher residues of pyrethroids in whole-rye bread can be explained by the fact that the purification step of the milling process is skipped.

Summarizing it can be said that the resulting pyrethroid residues are relatively stable during the half year storage and during the processing of the rye. It was observed that about 10–20 % of the pyrethroids being present in the rye cereal are to be expected to appear in the bread [119,120].

Pyrethroid residues corresponding to approx. 40 % of the original content were determined in bread prepared from whole crushed grain.

2.2.3.3 Wheat

In case of wheat only about 20 % of the pyrethroid residues get into the flour (Table 47, [119,120]). Wheat bread contains still about 20 or 10% resp., whole-wheat bread about

Table 46. Residues of pyrethroids in rye (grain protection and storage stability study)

Compound	Formulation	Days after last application	Country[1]	Residue levels (mg/kg) at harvest or at days after last application	Dosage a.i. (kg/ha)	Remarks (Ref.)
Cyfluthrin, Cypermethrin, Deltamethrin, Fenvalerate Permethrin	EC	0 days/months 0 days/10 months	D	0.2/0.2[a] 2.0/2.0[a]	0.2 mg/kg 2.0 mg/kg	[119] and [120]
		after milling; at the beginning of storage		0.10 in flour 0.25 in coarse flour 0.41 in bran 0.02–0.03 in bread 0.07–0.09 in whole-rye bread	0.2 mg/kg[b] of grain	

[1] D = Federal Republic of Germany.
[a] initial concentration of a.i. and concentration after a 10-months-storage-period;
[b] at the dosage level of 2 mg/kg the resulting residue values were about 10 times higher

30 or 50% resp. of the active ingredient content of the grain. No significant reduction of the residues results from toasting of wheat bread.

In case of wheat the degradation of pyrethroids during storage for half a year is very low. During milling about 20% of the residue remains in the flour fraction, the remainder is mainly in the bran [119,120].

After baking, the bread prepared from flour contained still about 10% and bread prepared from whole crushed grain about 50% of the original pyrethroid content of the grain [119,120].

In case of treatment of stored wheat with bioresmethrin in combination with the organic phosphates pirimiphosmethyl and/or fenitrothion, residues in the range of approx. 0.2 to 1 mg/kg resulted within 9 months after the application of 1 mg bioresmethrin/kg of wheat (Table 48 [125]).

The influence of different sotrage temperatures upon the residues of bioresmethrin on wheat during storage can be taken from Table 49 under [84]. The tests conducted in Australia show that the synergist piperonyl butoxide obviously has a strong influence upon the stability of the active ingredient, even at 35 °C, the highest storage temperature in these tests.

The results reported in Tables 50 and 51 under [124] originate from laboratory tests under controlled conditions. It can be seen that cyfluthrin degrades some-what more rapidly than cypermethrin. In case of both active ingredients the trans-isomers are degraded more rapidly than the cis-isomers. With both pyrethroids the diastereoisomer trans IV degraded most rapidly. One of the enantiomers of peak trans IV of cypermethrin seems to be degraded more rapidly than the other one. This means that an analysis being based only on the determination of the total pyrethroid content may be unreliable in case of cypermethrin as far as the efficacy of residues of this active gradient is concerned.

The rate of loss of both active ingredients is increased under conditions of higher storage temperatures and higher moisture contents of the cereals.

The breakdown of cyfluthrin and cypermethrin is relatively slow under normal storage conditions of the cereals.

The results reported in the Table 52 under [126] are based on an Australian study. In this case bulk wheat was stored in commercial silos. The duration of storage was 9 months. The pyrethroid residues on stored wheat are relatively persistent. An accumulation in the bran fraction is noticeable. On the other hand lower residues are present in white flour. The decline of the residues during the baking process is not significant.

The MRLs for cereal grains established by the Codex Alimentarius Commission in 1982 are: fenvalerate and phenothrin: 5 mg/kg; permethrin: 2 mg/kg; deltamethrin: 2 mg/kg (guideline level).

Given a moisture content of 10–11% which is the usual moisture content in Australian cereals, the residue situation of the 4 pyrethroids on wheat after storage can be regarded as safe.

Results from Australian studies concerning the stability of permethrin, phenothrin, fenvalerate and deltamethrin on wheat during storage are summarized in Table 53 under [128]. Supplementary to this Table the calculated half-lives of the substances are summarized. From these data it can be taken that a relatively slow degradation results which is accelerated, however, under conditions of higher temperatures and a higher

Table 47a. Residues of pyrethroids in wheat (grain protection and storage stability study)

Compound	Formulation	Storage (in weeks)	Country[1]	Residue levels (mg/kg) at harvest or at days after last application	Dosage a.i. (kg/ha)	Remarks (Ref.)
Cyfluthrin	EC	0/8/24/43 (at 10 °C storage temperature)	D	0.19/0.19/0.20/0.18	0.2 mg/kg of grain	119, 120)
		8/24/43 (at 20 °C storage temperature)		0.20/0.20/0.19		
Cypermethrin	EC	0/8/24/43 (at 10 °C storage temperature)		0.19/0.19/0.19/0.18	0.2 mg/kg of grain	119, 120)
		8/24/43 (at 20 °C storage temperature)		0.19/0.19/0.18		
Deltamethrin	EC	0/8/24/43 (at 10 °C storage temperature)		0.19/0.19/0.19/0.18	0.2 mg/kg of grain	119, 120)
		8/24/43 (at 20 °C storage temperature)		0.20/0.20/0.19		
Fenvalerate	EC	0/8/24/43 (at 10 °C storage temperature		0.19/0.18/0.19/0.18	0.2 mg/kg of grain	119, 120)
		8/24/43 (at 20 °C storage temperature)		0.19/0.19/0.19		
Permethrin	EC	0/8/24/43 (at 10 °C storage temperature)		0.19/0.18/0.19/0.17	0.2 mg/kg of grain	119, 120)
		8/24/43 (at 20 °C storage temperature)		0.19/0.19/0.18		

[1] D = Federal Republic of Germany

Table 47b. Residues of pyrethroids in wheat (Table cont.). Residue of five pyrethroids in the milling fractions of wheat after treatment (0.2 mg/kg) and after storage for 32 weeks

Duration of storage weeks	Milling fraction	Cyfluthrin mg/kg	Cypermethrin mg/kg	Deltamethrin mg/kg	Fenvalerate mg/kg	Permethrin mg/kg
0	flour (550)	0.040	0.042	0.037	0.041	0.046
0	coarse flour	0.31	0.31	0.29	0.31	0.31
0	bran	0.53	0.49	0.48	0.49	0.43
32	flour (550)	0.028	0.028	0.028	0.030	0.034
32	coarse flour	0.23	0.23	0.24	0.25	0.24
32	bran	0.51	0.52	0.57	0.54	0.44
Duration of storageweeks	**Milling fraction**	**Cyfluthrin %**	**Cypermethrin %**	**Deltamethrin %**	**Fenvalerate %**	**Permethrin %**
0	flour (550)	19	21	19	20	24
0	coarse flour	13	14	13	14	14
0	bran	68	65	68	66	62
32	flour (550)	14	14	13	15	19
32	coarse flour	11	11	10	11	12
32	bran	75	75	77	74	69

Table 47c. Residues of five pyrethroids in flour and whole crushed grain from wheat (treatment with 0.2 mg/kg, milling after storage for 13 weeks) and in the bread baked from these

Milling product or bread resp.	Cyfluthrin mg/kg	Cypermethrin mg/kg	Deltamethrin mg/kg	Fenvalerate mg/kg	Permethrin mg/kg
wheat flour (550)	0.032	0.034	0.031	0.035	0.039
white loaf	0.019	0.020	0.017	0.020	0.020
whole crushed wheat	0.17	0.17	0.16	0.16	0.17
whole-wheat bread	0.087	0.086	0.074	0.084	0.089

Table 48. Residues of bioresmethrin in pilot studies with pirimiphos-methyl plus bioresmethrin (code numbers 1 W–21 S) and fenitrothion plus bioresmethrin (code numbers 22 W–42 S), November 1975–August 1976 (according to J. Desmarchelier et al. [125])

Code number	Bioresmethrin residue ($mg\ kg^{-1}$) measured during										
	Nov.	Dec.	Jan.	Feb.	Mar.	Apr.	May	June	July	Aug.	Sept.
1W		1.0[b]	0.7		0.4	0.4				0.3	
2W				1.0[b]	0.8	0.8					
3W				1.0[b]	0.7	0.7					
5Q	1.0[b]				0.4						
6Q	1.0[b]				0.4					0.15	
7Q	1.1[b]				0.3				0.3	0.3	
8Q[a]	1.0[b]									0.25	
9Q	1.0[b]				0.3				0.15	0.15	
10Q		0.9[b]			0.3					0.15	
11N					1.2[b]	0.8		0.6			
12N					1.2[b]	0.7		0.6			
13N				1.2[b]	0.7						
14N				1.1[b]	0.9	0.6					
15N				1.1[b]	0.8	0.45					
16N				1.1[b]	0.8		0.65		0.5		0.35
17N				0.9[b]			0.65		0.5		0.35
18V			1.5[b]					0.5			
19V			1.0[b]								
20S[a]			0.9[b]							0.5	
21S			0.9[b]							0.2	
22W		0.9[b]	0.8		0.6					0.35	
23W		1.0[b]	0.8	0.7		0.65				0.5	
24W				1.1[b]		0.6				0.45	
25W				1.0[b]		0.65				0.5	
26Q	0.8[b]				0.3						
27Q	1.0[b]				0.4					0.25	
28Q	0.9[b]				0.3				0.1	0.1	
29Q[a]	1.4[b]				0.3					0.15	
30Q	1.0[b]								0.35		
31Q					0.5					0.25	
32N					1.0[b]	0.65		0.6			
33N					0.9[b]	0.5		0.45			
34N				0.7[b]	0.8	0.5					
35N				1.0[b]	0.9	0.45					
36N				0.8[b]	0.8	0.6					
37N				1.2[b]			0.65		0.3	0.35	
38N				1.1[b]			0.65		0.6	0.35	
39V			1,5[b]								
40V			1.2[b]				0.25				
41S[a]		0.8[b]								0.3	
42S		0.8[b]								0.15	

[a] Aerated;

[b] Calculated application rate; recommended application rate for bioresmethrin is 1 $mg\ kg^{-1}$. The calculation of the application rate was done by dividing the weight of the protectant in the applied spray by the weight of the grain. A 70-EC-formulation of the a.i. was used. The (commercial) storage conditions were: moisture 8.1–12.0% by weight.
Application rate: 1 g a.i./ton

Table 49. Bioresmethrin residues in wheat stored at various temperatures (according to WHO [84])

Treatment	Moisture content of grain %	Initial bioresmethrin level	Bioresmethrin residue, mg/kg, after storage at:				
			10 °C	20 °C	25 °C	30 °C	35 °C
Bioresmethrin on wheat	11.9	3.3	2.7	2.4	2.1	1.9	1.9
Bioresmethrin + piperonyl butoxide + antioxidant 1/10/1 on wheat	11.9	3.3	—	—	—	2.9	—
Bioresmethrin on wheat	10.2	2.8	—	—	2.3	—	2.2

moisture content of the wheat. In case of permethrin and phenothrin which consist of cis/trans-mixtures, the trans-isomer is degraded more rapidly than the cis-isomer.

Residues of cypermethrin and fenvalerate in milled fractions of stored wheat are reported under [129] in Table 54. It was determined that during baking of bread the residues are reduced by 12–21 % in comparison to those in the flour.

Cyfluthrin and cypermethrin were applied as grain protectants on bulk wheat in Australia (see Table 55 [135]). The highest residues were determined in bran and pollard. By baking bread the residues were reduced by approx. 29 %. After a storage for 9 months approx. 50–60 % of the initial residues of the two active ingredients were still detectable in the cereal grains.

Some operators reported of respiratory irritation while handling the substances. These effects may possibly be eliminated by the use of other formulation types of the two active ingredients. EC formulations were applied in the reported studies.

In Table 56 [127] results are summarized which were obtained after treatment of wheat for stored product protection with pyrethrins of natural origin. In this case the pyrethrins were applied as dust or liquid formulation into the grain stream as it goes into storage in order to prevent insect infestation. Two months after treatment an average amount of pyrethrins corresponding to approx. 50 % of the applied dosage was recovered. After storage for 12 months the recovery was approx. 25 %. This level remained until two years after application.

Table 57, Ref. [123], shows the residue situation on wheat herbage after treatment of *wheat plants* with cypermethrin and deltamethrin. A relatively rapid loss within the first hours after application was noted. Obviously the increase of plant matter plays only a minor role in the dissipation of the residues. In these tests no residues of the two active ingredients were detectable in the wheat (<0.01 mg/kg).

Various data from internal residue reports of Bayer AG on trials conducted with cyfluthrin on wheat in different countries are summarised in Table 58. These include also residue levels in processed products.

V. Kocourek [130] reported an interesting finding. He investigated the photostability of permethrin, deltamethrin and cypermethrin on treated wheat. These were laboratory tests with EC formulations. It was found that the time elapsing between treatment and UV irradiation is of importance. The results indicate that the pyrethroids can

Table 50. Cyfluthrin and cypermethrin on wheat in storage, Australia (according to R. M. Noble et al. [124]) Concentrations of the four diastereomers of cypermethrin on wheat stored for periods of up to 52 weeks at 25 or 35 °C with 12 or 15% moisture content

Initial concentration of pyrethroids (diastereoisomers) ($mg\ kg^{-1}$ of moist grain)		Storage conditions			Concentration of pyrethroid ($mg\ kg^{-1}$ of moist grain) after storage for following periods			
		Temperature (°C)	Moisture content (as% of wet weight)	Relative humidity (%)	13 weeks	26 weeks	39 weeks	52 weeks
Cis I	0.84, 0.81	25	12	54	0.75, 0.73	0.78, 0.74	0.69, 0.68	0.73, 0.71
Cis II	0.62, 0.61	25	12	54	0.55, 0.54	0.60, 0.57	0.49, 0.48	0.49, 0.48
Trans III	1.48, 1.45	25	12	54	1.39, 1.34	1.27, 1.21	1.18, 1.16	1.17, 1.13
Trans IV	1.00, 0.97	25	12	54	0.88, 0.85	0.76, 0.72	0.68, 0.66	0.58, 0.56
Cis I	0.81, 0.82	25	15	73	0.74, 0.70	0.76, 0.73	0.63, 0.62	0.65, 0.68
Cis II	0.60, 0.61	25	15	73	0.53, 0.50	0.55, 0.53	0.44, 0.43	0.41, 0.43
Trans III	1.44, 1.46	25	15	73	1.25, 1.19	1.19, 1.14	0.94, 0.92	0.87, 0.91
Trans IV	0.96, 0.98	25	15	73	0.79, 0.75	0.76, 0.72	0.57, 0.56	0.46, 0.48
Cis I	0.81, 0.82	35	12	57	0.74, 0.71	0.72, 0.75	0.58, 0.57	0.60, 0.59
Cis II	0.61, 0.61	35	12	57	0.53, 0.50	0.53, 0.55	0.43, 0.42	0.37, 0.37
Trans III	1.45, 1.46	35	12	57	1.22, 1.17	1.10, 1.15	0.85, 0.83	0.77, 0.76
Trans IV	0.97, 0.98	35	12	57	0.67, 0.64	0.55, 0.58	0.41, 0.40	0.34, 0.34
Cis I	0.82, 0.81	35	15	75	0.66, 0.68	0.69, 0.66	0.49, 0.52	0.45, 0.46
Cis II	0.60, 0.61	35	15	75	0.48, 0.49	0.50, 0.48	0.33, 0.36	0.28, 0.29
Trans III	1.46, 1.44	35	15	75	1.09, 1.11	1.01, 0.96	0.49, 0.53	0.57, 0.59
Trans IV	0.98, 0.97	35	15	75	0.62, 0.63	0.56, 0.53	0.21, 0.23	0.30, 0.31

Type of formulation used: cypermethrin (20% EC)

Table 51. Concentrations of the four diastereomers of cyfluthrin on wheat stored for periods of up to 52 weeks at 25 or 35 °C with 12 or 15% moisture content; according to R. M. Noble et al. [124]

Initial concentration of pyrethroids (diastereoisomers) (mg kg^{-1} of moist grain)		Storage conditions			Concentration of pyrethroid (mg kg^{-1} of moist grain) after storage for following periods			
		Temperature (°C)	Moisture content (as % of wet weight)	Relative humidity (%)	13 weeks	26 weeks	39 weeks	52weeks
Cis I	0.49, 0.49	25	12	54	0.42, 0.41	0.38, 0.37	0.37, 0.37	0.35, 0.35
Cis II	0.39, 0.39	25	12	54	0.34, 0.34	0.28, 0.27	0.29, 0.29	0.25, 0.26
Trans III	0.66, 0.66	25	12	54	0.56, 0.55	0.45, 0.44	0.38, 0.37	0.39, 0.40
Trans IV	0.47, 0.46	25	12	54	0.37, 0.37	0.31, 0.30	0.20, 0.20	0.21, 0.21
Cis I	0.53, 0.54	25	15	73	0.43, 0.40	0.36, 0.34	0.36, 0.35	0.36, 0.35
Cis II	0.42, 0.43	25	15	73	0.35, 0.32	0.29, 0.28	0.26, 0.25	0.27, 0.26
Trans III	0.71, 0.72	25	15	73	0.56, 0.51	0.41, 0.39	0.40, 0.39	0.33, 0.32
Trans IV	0.50, 0.51	25	15	73	0.39, 0.36	0.27, 0.26	0.18, 0.17	0.15, 0.15
Cis I	0.48, 0.49	35	12	57	0.41, 0.38	0.34, 0.30	0.30, 0.31	0.29, 0.29
Cis II	0.38, 0.39	35	12	57	0.32, 0.30	0.29, 0.25	0.23, 0.24	0.21, 0.20
Trans III	0.64, 0.66	35	12	57	0.49, 0.47	0.37, 0.33	0.28, 0.29	0.26, 0.25
Trans IV	0.45, 0.46	35	12	57	0.32, 0.30	0.20, 0.18	0.11, 0.11	0.15, 0.15
Cis I	0.48, 0.49	35	15	75	0.38, 0.37	0.31, 0.33	0.28, 0.27	0.23, 0.25
Cis II	0.38, 0.39	35	15	75	0.30, 0.29	0.25, 0.26	0.22, 0.22	0.18, 0.20
Trans III	0.65, 0.66	35	15	75	0.47, 0.46	0.33, 0.34	0.26, 0.30	0.20, 0.22
Trans IV	0.46, 0.46	35	15	75	0.29, 0.29	0.17, 0.18	0.12, 0.12	0.10, 0.11

Type of formulation used: cyfluthrin (5% EC)

Table 52a. Calculated insecticide application rates and the analytically determined residues on treated wheat stored in concrete silos at Natcha in Queensland and Noonbinna/Eugowra in New South Wales for varying times following spray application; according to M. Bengston [126)]

Treatment	Laboratory[a]	Site	Calculated application rate (mg/kg^{-1})	Insecticide residue[b] (mg kg^{-1}) at various times after treatment (months)						
				0.25	1.50	3	4.5	6	7	9
Deltamethrin (2 mg kg^{-1})	E	Natcha	2.2	1.55	1.67	1.53	1.61	1.53	—	1.42
+ piperonyl butoxide (8 mg kg^{-1})			8.6							
	E	Noonbinna	2.0	1.18	—	1.51	1.68	1.14	0.69	—
			8.0							
Fenitrothion (12 mg kg^{-1})	A	Natcha	10.9	9.2	6.2	5.8	4.9	4.4	—	4.8
+ fenvalerate (1 mg kg^{-1})	E		0.9	0.86	0.70	0.79	0.73	0.92	—	0.74
+ piperonyl butoxide (8 mg kg^{-1})			7.3							
	A	Eugowra	13.6	20.8	—	10.2	11.6	9.3	8.0	—
	E		1.1	1.16	—	1.01	0.79	0.80	0.80	—
			9.1							
Fenitrothion (12 mg kg^{-1})	A	Natcha	11.3	9.8	7.2	5.7	5.8	4.6	—	4.3
+ phenothrin (2 mg kg^{-1})	E		1.9	1.80	1.42	1.45	1.71	1.38	—	1.10
+ piperonyl butoxide (8 mg kg^{-1})			7.5							
	A	Noonbinna	11.1	—	—	7.8	8.2	9.0	7.4	0.7
	E		1.85	—	—	1.29	1.30	1.10	1.21	—
			7.7							
Pirimiphos-methyl (4 mg kg^{-1})	A	Natcha	4.0	2.3	2.3	2.1	1.8	1.9	—	1.8
+ permethrin (1 mg kg^{-1})	E		1.0	0.65	0.68	0.65	0.67	0.67	—	0.57
+ piperonyl butoxide (8 mg kg^{-1})			7.9							
	A	Eugowra	3.7	2.6	—	2.8	2.9	2.3	2.2	2.0
	E		0.93	0.70	—	—	0.69	0.65	0.60	0.56

[a] A, Entomology Branch; E, Agricultural Chemistry Branch, Queensland Department of Primary Industries;
[b] The data are the means from duplicate analyses of single subsamples. No dermination was made of piperonyl butoxide

Table 52b. Mean insecticide residues following the milling and baking of wheat stored in concrete silos at Natcha, Queensland for 10 months after application of the insecticides (Table cont.)

Insecticide	Insecticide residues ($mg\ kg^{-1}$)[a]					
	Wheat[b]	Bran	Pollard	Wholemeal bread	White flour	White bread
Deltamethrin	1.82	8.4	3.8	1.3	0.25	0.20
Fenvalerate	0.70	3.3	1.5	0.49	0.08	0.06
Permethrin	0.82	3.7	1.9	0.67	0.13	0.13
Phenothrin	1.21	4.4	3.6	0.81	0.23	0.20

[a] Laboratory E, Agricultural Chemistry Branch, Queensland Department of Primary Industries.
[b] Residues estimated from data on milled fractions.

Table 52c. Distribution of residues in milled fractions of treated wheat (Table cont.)

Fraction of treated wheat	Mean yield of fraction (%)	Distribution of residues (%)[a]			
		Deltamethrin	Fenvalerate	Permethrin	Phenothrin
Bran	14.4	66.5	68.5	61.7	54.1
Flour	74.6	10.3	8.5	12.0	14.1
Pollard	11.0	23.2	23.0	26.3	31.8

[a] Laboratory E, Agricultural Chemistry Branch, Queensland Department of Primary Industries

Table 52d. Mean insecticide residue levels in flour and in the bread baked from it, expressed on a moisture-free basis for both commodities (Table cont.)

Insecticide	Insecticide residues ($mg\ kg^{-1}$)[a]			
	White flour	White bread	Wholemeal flour	Wholemeal bread
Deltamethrin	0.29	0.32	2.1	2.2
Fenvalerate	0.09	0.10	0.80	0.73
Permethrin	0.15	0.19	0.93	1.0
Phenothrin	0.26	0.30	1.4	1.2

[a] Laboratory E, Agricultural Chemistry Branch, Queensland Department of Primary Industries.
EC-formulations of all a.i. were used. Rates of application: 2 mg/kg of deltamethrin and phenothrin and 1 mg/kg of fenvalerate and permethrin, respectively. In all cases 8 mg of piperonyl butoxide were added per kg. Initial moisture content of grain: 10–12%, temp.: 29–34 °C.

From Table 53 the following calculated half-lives (in weeks) result

Compound	25 °C/12% moisture	35 °C/15% moisture
Phenothrin	72	29
Deltamethrin	114	35
Fenvalerate	210	74
Permethrin	252	44

Table 53. Concentration of four pyrethroids found on wheat stored for periods of up to 52 weeks at 25 or 35 °C with 12 or 15% moisture content (according to R. M. Noble et al. [128])

Initial concentration of pyrethroid (replicates) (mg kg^{-1} of moist grain)		Storage conditions			Concentration of pyrethroid (mg kg^{-1} of most grain) after storage for following periods			
		Temperature (°C)	Moisture content (as % of wet weight)	Relative humidity (%)	13 weeks	26 weeks	39 weeks	52 weeks
Permethrin	0.96, 0.92	25	12	54	1.00	0.89, 0.89	0.81	0.87, 0.91
	1.03, 1.06	25	15	73	1.03	0.87, 0.86	0.75	0.83, 0.77
	0.95, 0.98	35	12	57	0.91	0.77, 0.77	0.60	0.74, 0.68
	1.02, 0.97	35	15	75	0.97	0.64, 0.60	0.57	0.52, 0.37
Phenothrin	2.11, 1.73	25	12	54	1.79	1.62, 1.64	1.56	1.07, 1.15
	2.0, 2.0	25	15	73	1.83	1.66, 1.70	1.24	1.01, 0.86
	1.92, 1.90	35	12	57	1.40	1.21, 1.30	1.08	0.72, —
	2.07, 1.97	35	15	75	1.51	1.10, 1.17	0.96	0.54, 0.53
Fenvalerate	1.05, 1.16	25	12	54	1.13	1.19, 1.13	1.13	0.86, 0.90
	1.14, 1.08	25	15	73	1.16	1.19, 1.22	1.14	0.88, 0.81
	1.18 —	35	12	57	1.08	1.17, 1.05	0.96	0.71, 0.77
	1.14 —	35	15	75	1.06	0.87, 1.06	0.87	0.64, 0.68
Deltamethrin	2.07 —	25	12	54	1.91	1.89, 1.77	1.66	1.51, 1.42
	2.13 —	25	15	73	2.00	2.00, 1.92	1.64	1.35, 1.28
	2.01 —	35	12	57	1.83	1.67, 1.66	1.38	1.17, 1.26
	1.95 —	35	15	75	1.74	1.39, 1.29	1.25	0.62, 0.70

Types of formulation: deltamethrin: 25 EC; fenvalerate: 50 EC; permethrin: 500 EC; phenothrin: 100 EC

Table 54. Cypermethrin and fenvalerate residue a levels[a] in white and wholemeal flour and bread (expressed on a moisture-free basis); according to B. S. Joia [129]

Insecticides	Residues, mg kg^{-1}			
	White flour	White bread	Wholemeal flour	Wholemeal bread
Cypermethrin	0.51	0.43 (84.3)[b]	3.50	2.78 (79.4)
Fenvalerate	0.66	0.58 (87.9)	3.96	3.44 (86.9)

[a] Mean of three replications; [b] Percent residues remaining after baking.
Rate of application: 8 and 12 mg a.i./kg of wheat grain;
Moisture content: 13.3 and 15.0%;
Storage conditions: 60 weeks at −5 °C and at 25 °C

penetrate into the interior of the grain. For the investigation of such penetration effects the technique of UV treatment is suitable.

This chapter shall be concluded with a general hint. Especially in the storage of wheat (this applies of course also to other cereals and grain products in storage having been discussed or still to be discussed), the longer residual effect of pyrethroids especially under conditions of higher storage temperatures and higher degrees of moisture is a very favourable property in those countries where such conditions prevail. That is to say that the organic phosphate insecticides being applicable as an alternative degrade more rapidly under such conditions. A consequence of this are frequent and expensive applications with the selection pressure for resistance increasing as a consequence of low levels of fast degrading insecticides.

2.2.3.4 Corn

The following is to be added concerning Table 59 on residues of pyrethroids in corn: according to [72] tefluthrin was used as soil insecticide in corn. It was detected that the parent compound is taken up by the plant through the roots into the foliage (leaves and stem). On the other hand, no residues of tefluthrin were detectable in the grains. Of interest is the finding that in this case the residues consist mainly of free and conjugated degradation products of the substance, while residue of the parent compound tefluthrin amount to only a portion of 1–8% of the total residues.

The residue data on deltamethrin and permethrin in corn [131] summarized in the same Table are taken from a publication which is based on investigations concerning the persistence of insecticidal spray deposits on woven polypropylene fabrics and the transport of the a.i. into corn.In these investigations it was found that WP formulations result in a greater persistence of the parent compounds on the surface and lower residues in corn than EC formulations.

Table 60 shows some residue data of cyfluthrin in corn with residue levels in cobs and kernels being below the limit of determination of 0.04 mg/kg.

2.2.3.5 Rice

In Table 61 residue data concerning pyrethroids on rice are summarized. The paper of Desmarchelier [118] is based on the use of bioresmethrin and d-fenothrin in stored

Table 55a. Calculated insecticide application rates and the analytically determined residues on treated wheat stored in concrete silos at Malu in Queensland and Garema in New South Wales for varying times following spray application (according to M. Bengston et al [135])

Treatment	Site	Laboratory[a]	Calculated application rate (mg kg^{-1})	Insecticide residue[b] (mg kg^{-1}) at various times after treatment (months)					
				0	$1^1/_2$	3	$4^1/_2$	6	9
Cyfluthrin (2 mg kg^{-1})	Malu	E	1.90	1.6	1.4	1.3	1.4	1.2	1.1
+ piperonyl butoxide (10 mg kg^{-1})		F		1.8	1.5	1.4	1.4	1.3	1.1
	Garema	E	1.74	1.1	0.76	1.6	1.3	0.8	—
		F		1.2	0.74	1.6	1.3	1.1	1.1
Cypermethrin (4 mg kg^{-1})	Malu	E	4.20	3.0	3.1	2.5	2.4	2.2	2.2
+ piperonyl butoxide (10 mg kg^{-1})		G		2.4	2.6	2.6	3.0	2.8	2.8
	Garema	E	4.59	3.1	3.1	3.0	2.6	2.4	—
		G		3.0	2.6	2.4	3.4	3.0	—

[a] Laboratory E — Agricultural Chemistry Branch, Queensland Department of Primary Industries; Laboratory F — Bayer AG, Leverkusen, Federal Republik of Germany; Laboratory G — Shell Chemical (Australia) Proprietary Limited;

[b] The data are the means from duplicate analyses of single subsamples. No determination was made of piperonyl butoxide

Table 55b. Mean insecticide residues following milling and baking of wheat stored in concrete silos at Malu, Queensland for 9 months after application of the insecticides (Table cont.)

Sample analysed	Laboratory[a]	Cyfluthrin ($mg\ kg^{-1}$)	Cypermethrin ($mg\ kg^{-1}$)
Wheat	E	1.3 (1.4[b])	2.1 (2.3[b])
	F	1.1	—
Bran	E	4.3 (4.9[b])	10.0 (11.6[b])
	F	3.5	—
Pollard	E	2.9 (3.3[b])	5.3 (6.0[b])
		—	—
White flour	E	0.18 (0.21[b])	0.30 (0.35[b])
White bread	E	0.11 (0.16[b])	0.18 (0.26[b])
	F	0.08	—
Wholemeal flour	E	1.1 (1.3[b])	2.1 (2.4[b])
		—	—
Wholemeal bread	E	0.84 (1.3[b])	1.4 (2.1[b])

[a] Laboratory E — Agricultural Chemistry Branch, Queensland Department of Primary Industries; Laboratory F — Bayer AG, Leverkusen, Federal Republic of Germany;
[b] Moisture-free basis

Table 55c. Distribution of residues in milled fractions of treated wheat (Table cont.)

Fraction of treated wheat	Mean yield of fraction (%)	Residue (% of total)	
		Cyfluthrin[a]	Cypermethrin[a]
Bran	12.5	52.0	62.9
Flour	77.0	14.3	11.1
Pollard	10.5	33.7	26.0

[a] Laboratory E — Agricultural Chemistry Branch, Queensland Department of Primary Industries

product protection. Additionally the fate of residues during processing of rice was investigated. Milling of rice was the main cause for the reduction of the residue levels. After the milling process, residues of the two pyrethroids applied to the unhusked rice were concentrated in the husks and bran.

From Table 62 [84)] it can be taken that piperonyl butoxide obviously influence the stability of the bioresmethrin residues, even at a storage temperature of 35 °C. The biological half-lives are temperature-dependent, which is to be expected. At 35 °C the half-life is about 8 weeks, at 20 °C it is above 20 weeks.

The residues cited in Table 63 under [132)] are from experiments conducted in a Japanese study concerning the reduction of the photodegradation of pyrethrins which were encapsulated in beta-cyclodextrine. Besides the pyrethrins the photodegradation-reducing action of the beta-cyclodextrine was also successfully tested with resmethrin.

Table 55d. Percentage reduction in insecticide residue (wet weight basis) during storage of treated grain in silos and following milling and baking of subsamples (Table cont.)

Percentage loss	Laboratory[a]	Reduction in residues (% of initial)			
		Cyfluthrin		Cypermethrin	
		Malu, Qld	Garema, NSW	Malu, Qld	Garema, NSW
During storage[b]	E	31.3	27.3	26.7	22.6
	F	38.9	8.3	—	—
	G	—	—	−16.7[c]	0
From raw grain to:					
wholemeal flour	E	15.4	—	0	—
white flour	E	86.2	—	85.7	—
wholemeal bread	E	35.4	—	33.3	—
white bread	E	91.5	—	91.4	—

[a] E — Agricultural Chemistry Branch, Queensland Department of Primary Industries; F — Bayer AG, Leverkusen, Fed. Rep. of Germany; G — Shell Chemical (Australia) Proprietary Limited.

[b] Calculated by comparing residue of final sample with residue of the initial sample.

[c] Residues in initial sample were unexpectedly low.

The results reported in Table 55 are according to Ref. [135]. The initial moisture content of the grain ranged between 8.6 and 11.6%. All wheat used was graded Australian Standard White. The quantity of grain treated ranged between 510 and 864 tonnes.

For further details see text

Table 56. Percentage of recovery of pyrethrins and piperonyl butoxide from treated wheat after aging for 2, 12 or 24 months (according to H. H. Walkden [127])

Treatment	Percent recovery after aging (months)		
	2	12	24
1.5 mg/kg of pyrethrins	40	2	40
2.0 mg/kg of pyrethrins	59	30	25
2.5 mg/kg of pyrethrins	54	28	28
25.0 mg/kg of piperonyl butoxide	83	61	58
1.36 mg/kg of pyrethrins	46	30	Trace
1.36 mg/kg of pyrethrins with MGK-264, 1-to-5 ratio[a]	40	30	—
1.36 mg/kg of pyrethrins with MGK-264, 1-to-10 ratio	20	22	—
1.36 mg/kg of pyrethrins	26	22	15

[a] MGK-264 is used as a synergist (= N-octyl-bicycloheptene dicarboximide)

Table 57. Relative plant growth on dry weight basis, concentration (mg/kg), standard deviation (s.d.), and percent of initial concentration of deltamethrin and cypermethrin on *wheat herbage* at seven intervals after application (according to N. D. Westcott [123])

Time	Relative Growth	Deltamethrin mg/kg (s.d.)	(%)	Cypermethrin mg/kg (s.d.)	(%)
0.2 hrs	1.00	0.70 (0.12)	100	3.74 (0.25)	100
3 hrs	1.00	0.49 (0.09)	70	2.63 (0.40)	70
10 hrs	1.00	0.32 (0.02)	46	2.07 (0.46)	55
1 day	1.05	0.32 (0.12)	46	1.89 (0.14)	50
3 days	1.13	0.24 (0.04)	34	1.52 (0.21)	40
6 days	1.10	0.16 (0.06)	23	1.24 (0.12)	33
27 days	1.25	0.05 (0.02)	7	0.20 (0.03)	5

s.d. = standard deviation;
Rate of application: cypermethrin: 28 g a.i./ha (EC-formulation);
deltamethrin: 6 g a.i./ha (EC-formulation)

2.2.3.6 *Sorghum*

Sorghum was treated during storage with mixtures of pyrethroids (bioresmethrin, (1 R)-phenothrin and pyrethrins) and organophosphates. Partly piperonyl butoxide was added as synergist for the pyrethroids. The resulting pyrethroid residue levels are summarized in Table 64 under [133].

In general pyrethroids are suggested to replace the very widely used malathion for stored product protection in cereals, because the respective insect pests have already developed resistance to malathion in many cases.

The results reported under [134] in Table 64 are also based on tests with sorghum in stored product protection. Also in this case mixtures of insecticidal organophosphates with the pyrethroids deltamethrin, fenvalerate, permethrin and phenothrin were used, the latter again synergized with piperonyl butoxide. In this article [134] it was only reported that in all cases the residues were below the MRLs established by the Codex Alimentarius Commission.

2.2.4 Various Commodities

2.2.4.1 *Oilseed and Oil Producing Plants*

Table 65 shows residue data of cyfluthrin in rape with the residue levels in pods and seeds being at or below the respective limits of determination.

The residue situation of pyrethroids on/in cottonseed including lint is summarized in Table 66.

In addition to the presented data and for their evaluation the following should be mentioned. Detailed information concerning the test conditions are missing with the results given under [83]. This is to be attributed to the fact that this publication is basically a method report.

The results reported under [136] were obtained after 4 applications of the four active ingredients flucythrinate, deltamethrin, cypermethrin and fenvalerate. The first of these four applications was made when 50% of the flower setting was initiated, the

Table 58. Residues of cyfluthrin in wheat (Internal Residue Reports, Bayer AG, Leverkusen, FRG)

Formulation	Days after last application	Country[1]	Residue levels (mg/kg) at harvest or at days after last application	Dosage a.i. (kg/ha)
50 SL	0/46/95/137/179/259	AUS	1.9/1.8/1.5/1.4/1.3/1.1 (grains)	1 × 2 g/t
			bread 0.08 (259 days)	1 × 2 g/t
			dark bread 1.4 (259 days)	1 × 2 g/t
			bran 3.5 (259 days)	1 × 2 g/t
			flour 0.12 (259 days)	1 × 2 g/t
50 SL	1/52/105/135/189/233	AUS	1.2/0.74/1.7/1.3/1.1/1.1 (grains)	1 × 1.6 g/t
50 EC	62	S	<0.01	1 × 0.025
50 EC	0/53/89/129/171/255/270	AUS	grains 1.2/1.2/1.1/1.1/1.1/1.0/0.86	1 × 2 g/t
			flour 0.15 270 days	1 × 2 g/t
			wholemeal 0.95 270 days	1 × 2 g/t
			white bread 0.08 270 days	1 × 2 g/t
			wholemeal bread 0.62 270 days	1 × 2 g/t
			bran 5.10 270 days	1 × 2 g/t
			pollard 2.20 270 days	1 × 2 g/t
10 WP	1/56/100/143/191/218	AUS	grains 1.1/1.2/0.98/1.1/1.1/0.98	1 × 1.95 g/t
	218		flour 0.23	1 × 1.95 g/t
	218		wholemeal 0.87	1 × 1.95 g/t
	218		white bread 0.43	1 × 1.95 g/t
	218		wholemeal bread 0.51	1 × 1.95 g/t
	218		bran 2.60	1 × 1.95 g/t
	218		pollard 2.60	1 × 1.95 g/t
50 EC	22/28	BR	<0.01/<0.01 (grains)	1 × 0.025
50 EW	1/43/85/127/169	AUS	0.92/ 0.82/0.68/0.96/0.76 } grains	1 × 2 g/t
50 EW	2/43/85/134	AUS	1.43/ 0.47/0.66/0.54 } grains	1 × 2 g/t
50 EC	72	S	<0.04 } grains	1 × 0.025
50 EW	0/30/61/91/122/183	RA	2.50/ 1.40/1.60/1.00/1.20/0.90 grains	1 × 2 g/t
	214/244/275	RA	1.00/ 0.80/0.94 grains	1 × 2 g/t
50 EW	0/30/61/91/122/159	RA	4.30/ 3.71/2.10/3.53/2.83/2.73 grains	1 × 2 g/t
	183/214/244/275	RA	2.33/ 2.50/2.47/2.22 grains	1 × 2 g/t
50 EC	3	E	0.05 (grains)	1 × 0.025
50 EC	3	E	2.94 (straw)	1 × 0.025

100 EW	283	<0.01 (grains and straw each)	1 × 6 g/100 kg seed
100 EW	283	<0.01 (grains and straw each)	1 × 6 g/100 kg seed
100 ES	160	<0.01 (grains and straw each)	1 × 15 g/100 kg seed
100 EW	154	<0.01 (grains and straw each)	1 × 15 g/100 kg seed
100 ES	151	<0.01 (grains and straw each)	1 × 15 g/100 kg seed
100 ES	147	<0.01 (grains and straw each)	1 × 15 g/100 kg seed

[1] AUS = Australia; BR = Brazil; E = Spain; RA = Argentina; S = Sweden

Table 59. Residues of pyrethroids in corn

Compound	Formulation	Days after last application	Country[1]	Residue levels (mg/kg) at harvest or at days after last application	Dosage a.i. (kg/ha)	Remarks (Ref.)
Tefluthrin	Granul.	at harvest (82—55 days after appl.)	UK	0.01—0.46 (leaves and stem)[a] <0.01 (grain)	0.8 and 0.3 resp.	72)
Deltamethrin	EC or WP	0/12 weeks 0/12 weeks	UK	<0.05/<0.05 in grain <0.05/<0.05	6.2 mg/m²	131b) 12.5 mg/m²
Permethrin	EC or WP	0/12 weeks 0/12 weeks		<0.1/<0.1 in grain <0.1/<0.1	41 mg/m² 83 mg/m	

[1] UK = United Kingdom;
[a] residues after soil application of ^{14}C-a.i., values calculated as cyclopropane- and phenyl-^{14}C-residues; [b] persistence study of spray deposits on woven polypropylene fabric (details: see text)

Table 60. Residues of cyfluthrin in corn (Internal Residue Reports, Bayer AG, Leverkusen, FRG)

Formulation	Days after last application	Country[1]	Residue levels (mg/kg) at harvest or at days after last application	Dosage a.i. (kg/ha)
50 EC	52	F	<0.01	1 × 0.025
50 EC	52	F	<0.01	1 × 0.025
10 WP	184	BR	0.55	1 × 2 ppm (direct treatment of kernels)
10 WP	184	BR	0.68	1 × 2 ppm (direct treatment of kernels)
10 WP	184	BR	1.80	1 × 4 ppm (direct treatment of kernels)
10 WP	184	BR	1.70	1 × 4 ppm (direct treatment of kernels)
50 EC	0/14/21	D	1.00/ 0.07/0.04	1 × 0.038 whole plant
50 EC	28/35/42	D	0.04/ 0.04/0.04	1 × 0.038 green mass
50 EC	28/35	D	<0.04/<0.04	1 × 0.038 cobs
50 EC	42/68	D	<0.04/<0.04	1 × 0.038 kernels
50 EC	0/14/21	D	0.87/ 0.16/0.11	1 × 0.038 whole plant
50 EC	28/35/42	D	0.07/ 0.07/0.05	1 × 0.038 green mass
50 EC	28/35	D	<0.04/<0.04	1 × 0.038 cobs
50 EC	42/76	D	<0.04/<0.04	1 × 0.038 kernels
50 EC	0/14/21	D	0.80/ 0.35/0.13	1 × 0.038 whole plant
50 EC	28/35	D	<0.04/<0.04	1 × 0.038 green mass
50 EC	28/35	D	0.06/<0.04	1 × 0.038 cobs
50 EC	84	D	<0.04	1 × 0.038 kernels
50 EC	84	D	<0.04	1 × 0.038 straw
50 EC	0/14/21	D	0.081/ 0.08/0.06	1 × 0.038 whole plant
50 EC	28/35	D	0.06/<0.04	1 × 0.038 green mass
50 EC	28/35	D	<0.04/<0.04	1 × 0.038 cobs
50 EC	84	D	<0.04	1 × 0.038 kernels
50 EC	84	D	<0.04	1 × 0.038 straw
50 EW	117	D	<0.04	1 × 0.050 kernels
50 EW	103	D	<0.04	1 × 0.050 kernels
25 SL	82	D	<0.04	1 × 0.025 kernels
25 SL	82	D	<0.04	1 × 0.025 kernels

50 EC	0/14/21/28/35	D	1.40/ 0.07/ 0.05/ 0.05/ <0.04		1 × 0.038 whole plant
50 EC	28/35	D	<0.04/ <0.04		1 × 0.038 cobs
50 EC	77	D	<0.04		1 × 0.038 kernels
50 EC	0/14/21/28/35	D	0.50/ 0.08/ 0.09/ 0.05/ 0.05		1 × 0.038 whole plant
50 EC	28/35	D	<0.04/ <0.04		1 × 0.038 cobs
50 EC	77	D	—0.04/		1 × 0.038 kernels
50 EC	0/14/21/28/35	D	0.78/ 0.08/ <0.04/ <0.04/ <0.04		1 × 0.038 whole plant
50 EC	28/35	D	<0.04/ <0.04		1 × 0.038 cobs
50 EC	93	D	<0.04		1 × 0.038 kernels
50 EC	0/14/21/28/35	D	0.94/ 0.08/ 0.04/ 0.04/ <0.04	= = = = =	1 × 0.038 whole plant
50 EC	28/35		<0.04/ <0.04		1 × 0.038 cobs
50 EC	93		<0.04		1 × 0.038 kernels
50 EC	3	E	<0.04		1 × 0.025 kernels
50 EC	15	E	<0.04		1 × 0.025 kernels

[1] F = France; D = Fed. Rep. of Germany; E = Spain; BR = Brazil

Table 61 a. Residues of pyrethroids on husked rice (relative humidity 55 %) after storage at 25 °C and after subsequent cooking; according to J. M. Desmarchelier et al. [118]

Protectant	Application level (mg/kg)	Residues (mg/kg)		Residues (mg/kg)[a] after cooking	
		After 3 months	After 6 months	At 3 months	At 6 months
Bioresmethrin	7	4.5	4.0	2.0	2.0
d-Fenothrin	8	—	6.2	—	4.1

[a] Residues expressed as if on rice, 55 % R. H.

Table 61 b. Residues of pyrethroids on polished rice (relative humidity 55 %) after storage at 25 °C and after subsequent cooking

Protectant	Application level (mg/kg)	Residues (mg/kg)		Residues (mg/kg)[a] after cooking	
		After 3 months	After 6 months	At 3 months	At 6 months
Bioresmethrin	7	5.1	3.5	2.0	0.8
d-Fenothrin	8	—	5.9	—	3.1

[a] Residues expressed as if on rice, 55 % R. H.

Table 61 c. Residues of pyrethroids on unhusked rice (relative humidity 65 %) after storage at 25 °C and after subsequent milling and cooking of milled products

Protectant	Application levels (mg/kg)	Residues (mg/kg)		Residues (mg/kg) after 6 months storage and			
		After 3 months	After 6 months	Milling		Milling and cooking[a]	
				Husked rice	Polished rice	Husked rice	Polished rice
Bioresmethrin	7	4.5	3.5	0.25	0.1	0.05	—0.05
d-Fenothrin	8	—	4.7	0.50	0.15	0.2	<0.1

[a] Residues expressed as if on rice, 65 % R. H.
Formulations: 200 EC (Bioresmethrin) and 800 EC (d-Fenothrin); Country: Australia

Table 62. Bioresmethrin residues in rice stored at various temperatures

Treatment	Moisture content of grain %	Initial bioresmethrin level	Bioresmethrin residue, mg/kg, after storage at:				
			10 °C	20 °C	25 °C	30 °C	35 °C
Bioresmethrin on husked rice	12.0	3.3	2.9	2.3	—	1.5	—
Bioresmethrin + piperonyl butoxide (1/10) on husked rice	12.0	3.3	—	3.1	—	2.9	—
Bioresmethrin on polished rice	13.0	2.8	—	—	0.8	—	0.8
Bioresmethrin + piperonyl butoxide (1/10) on polished rice	13.0	2.8	—	—	1.7	—	1.4

Table 62 is taken from WHO (84); Formulation: not given; Country: Australia

Table 63. Residues of pyrethroids in rice

Compound	Formulation	Days after last application	Country[1]	Residue levels (mg/kg) at harvest or at days after last application	Dosage a.i. (kg/ha)	Remarks (Ref.)
Pyrethrins	"Dust or WP"	5/7/14	JAP	<0.03/<0.03/<0.03 (husked rice)	0.1 g a.i./l (6 applicat.)	132)
		5/7/14		0.12/ 0.07/ 0.05 (straw)		

[1] JAP = Japan

following applications were made at 15 day intervals thereafter. The last application was made three weeks prior to the first sampling.

Recommended MRLs for fenvalerate and cypermethrin are 0.2 and 0.1 mg/kg respectively [136)]. All pyrethroids mentioned can therefore be considered as safe on this crop.

The results reported in Table 66 under [137)] originate from a study for the investigation of the translocation of fluvalinate residues from cotton plants into the seed. The applications started at the time of boll formation and were continued at 11–15 day intervals. 20–25 days after the last spray, open bolls were picked for the determination of the residues. The residues in seed are markedly below the MRL of 0.2 mg/kg [137)].

According to [138)] the MRL for permethrin is 0.5 mg/kg, the respective MRLs for cypermethrin and fenvalerate were already cited under [136)].

The results listed in Table 66 under [139)] show that rates of 50 or 100 g of permethrin per ha can be considered as safe with regard to the residue situation.

Table 64. Residues of pyrethroids on sorghum

Compound	Formulation	Days after last application	Country[1]	Residue levels (mg/kg) at harvest or at days after last application	Dosage a.i. (kg/ha)	Remarks (Ref.)
Bioresemthrin (1R)-Fenothrin and Pyrethrins	EC	0—24 weeks	AUS	between 2 and 0.4	1 mg/kg, 1 mg/kg and 1.5 or 2 mg/kg	133)
Deltamethrin, Fenvalerate, Permethrin and Phenothrin	n.g.	at harvest	AUS	<2[a] <5 <2 <5	2 mg/mg, 1 mg/kg, 1 mg/kg and 2 mg/kg	134)

[1] AUS = Australia; n.g. = not given;
[a] the values given are within the residue limits established by the Codex Alimentarius Commission (for further details: see text);
In exp. 134) the mean moisture content of grain was 12.0%, the mean grain temperature was 22.8 °C. All the pyrethroids were synergized with piperonyl butoxide (8 mg/kg)

Table 65. Residues of cyfluthrin in rape (Internal Residue Reports, Bayer AG, Leverkusen, FRG)

Formulation	Days after last application	Country[1]	Residue levels (mg/kg) at harvest or at days after last application	Dosage a.i. (kg/ha)
50 EC	241	F	<0.05	1 × 0.015
50 EC	260	F	<0.05	1 × 0.015
50 EC	68	S	<0.05	1 × 0.025 rapeseed
50 EC	65	S	<0.05	1 × 0.025
50 EC	0/42	D	0.13/ 0.02	2 × 0.015 rape (green matter)
50 EC	42/49/56	D	<0.10/<0.10/<0.10	2 × 0.015 pods
50 EC	72	D	<0.05	2 × 0.015 seeds
50 EC	0/42/49/59	D	0.15/<0.02/<0.02/<0.02	2 × 0.015 green matter
50 EC	42/49/56	D	<0.10/<0.10/<0.10	2 × 0.015 pods
50 EC	73	D	<0.05	2 × 0.015 seeds
50 EC	0	D	0.20	2 × 0.015 green matter
50 EC	42/49/56/63	D	0.03/ 0.04/ 0.02/ 0.02	2 × 0.015 stems
50 EC	42/49/56/63	D	<0.10/<0.10/<0.10/<0.10	2 × 0.015 pods
50 EC	70	D	0.05	2 × 0.015 seeds
50 EC	0	D	0.34	2 × 0.015 green matter
50 EC	42/49/56/63	D	<0.02/ 0.02/ 0.02/<0.02	2 × 0.015 stems
50 EC	42/49/56	D	<0.10/<0.10/<0.10	2 × 0.015 pods
50 EC	63	D	<0.05	2 × 0.015 seeds
50 EC	89	F	<0.05	1 × 0.01
50 EC	89	F	<0.05	1 × 0.01 seeds
50 EC	0/42/49	D	0.32/<0.02/<0.02	2 × 0.015 green matter
50 EC	56	D	0.06	2 × 0.015 straw
50 EC	42/49/56	D	<0.10/<0.10/<0.10	2 × 0.015 pods
50 EC	66	D	0.05	2 × 0.015 seeds
50 EC	0/42/49/56	D	0.27/<0.02/<0.02/<0.02	2 × 0.015 green matter
50 EC	42/40/56	D	<0.10/<0.10/<0.10	2 × 0.015 pods
50 EC	75	D	<0.05	2 × 0.015 seeds
50 EC	0	D	0.21	2 × 0.015 green matter
50 EC	42/49/56	D	<0.02/<0.02/<0.02	2 × 0.015 stems
50 EC	42/49	D	<0.10/<0.10	2 × 0.015 pods
50 EC	56	D	<0.05	2 × 0.015 seeds
50 EC	0	D	0.18	2 × 0.015 green matter
50 EC	21/28	D	0.06/ 0.04	2 × 0.015 whole plants
50 EC	21/28	D	<0.10/<0.10	2 × 0.015 pods
50 EC	35	D	<0.05	2 × 0.015 seeds

[1] F = France; S = Sweden; D = Federal Republik of Germany

The results reported in Table 66 under [140] are based on 5 applications at 15-day intervals, the sampling was made 3 weeks after the last application.

Concerning the residue data presented under [142] it should be noted that flucythrinate has a MRL of 0.1 mg/kg on cotton seed (EPA, 1982). Flucythrinate can be safely recommended to cotton growers.

Table 66. Residues of pyrethroids in cottonseed

Compound	Formulation	Days after last application	Country[a]	Residue levels (mg/kg) at arvest or at days after last application	Dosage a.i. (kg/ha)	Remarks (Ref.)
Fenvalerate	n.g.	77	ISR	0.012	2 × 0.3	83)
				0.053	2 × 0.6	
Flucythrinate	10 EC	21	IND	seed: ND[b]/lint: ND	4 × 0.03	136)
				seed: ND /lint: ND	4 × 0.05	
Deltamethrin	2.8 EC	21	IND	seed: ND /lint: ND	4 × 0.01	
				seed: ND /lint: ND	4 × 0.015	
Cypermethrin	10 EC	21	IND	seed: 0.05 /lint: 0.15	4 × 0.04	
				seed: 0.08 /lint: 0.25	4 × 0.075	
Fenvalerate	20 EC	21	IND	seed: ND /lint: 0.02	4 × 0.04	
				seed: 0.06 /lint: 0.20	4 × 0.075	
Fluvalinate	25 EC	20—25	IND	seed: 0.05 /lint: 0.35	4 × 0.05[e]	137)
				seed: 0.06 /lint: 0.90	4 × 0.10[e]	
				seed: ND /lint: 0.85	4 × 0.06[f]	
				seed: ND /lint: 0.61	3 × 0.05[g]	
				seed: 0.04 /lint: 0.90	3 × 0.10[g]	
Cypermethrin	EC	at harvest	IND	seed: 0.08 /lint: 0.88	6 × 0.06	138)
Fenvalerate	EC	at harvest	IND	seed: 0.10 /lint: 1.66	6 × 0.12	
Permethrin	EC	at harvest	IND	seed: 0.16 /lint: 0.44	6 × 0.09	
Permethrin	25 EC	at harvest	IND	seed: ND[d]/lint: ND	2 × 0.05	139)
				seed: ND /lint: ND	2 × 0.10	
Cypermethrin	25 EC	21	IND	seed: ND[c]/lint: 0.123	5 × 0.05	140)
				seed: ND /lint: 0.243	5 × 0.075	and
				seed: 0.002/lint:0.284	5 × 0.10	141)
Deltamethrin	2.5 EC	21	IND	seed: ND /lint: 0.009	5 × 0.01	
				seed: ND /lint: 0.062	5 × 0.015	
				seed: ND /lint: 0.167	5 × 0.020	

Fenvalerate	20 EC	21	IND	seed: ND /lint: 0.008	5 × 0.05	
				seed: 0.03 /lint: 0.014	5 × 0.075	
				seed: 0.04 /lint: 0.05	5 × 0.10	
Permethrin	50 EC	21	IND	seed: ND /lint: 0.016	5 × 0.05	
				seed: ND /lint: 0.042	5 × 0.075	
				seed: 0.04 /lint: 0.216	5 × 0.10	
Flucythrinate	25 EC	at harvest	IND	seed: ND[b]/lint: 0.15	3 × 0.06[f]	142)
				seed: ND /lint: 0.24	3 × 0.05[g]	
				seed: 0.05 /lint: 0.41	3 × 0.10[g]	

[a] India;
[b] ND = not detectable (<0.02 mg/kg);
[c] ND = not detectable (<0.001 ppm);
[d] ND = not detectable (limit of determination not given);
[e] Experiments in 1983;
[f] Experiments in 1984;
[g] Experiments in 1985;
n.g. = not given

Table 67. Residues of cyfluthrin in cotton (Internal Residue Reports, Bayer AG, Leverkusen, FRG)

Formulation	Days after last application	Country[1]	Residue levels (mg/kg) at harvest or at days after last application	Dosage a.i. (kg/ha)
50 SL	15	IND	<0.03	4 × 0.025
50 SL	25	IND	<0.03	4 × 0.025
50 SL	36	ET	<0.03	4 × 0.09
100 EC	7	T	<0.03	total 0.217
100 EC	7	T	0.04	total 0.217
100 EC	7	T	<0.03	total 0.217
50 SL	14	ET	<0.03	4 × 0.09
50 EC	18	GCA	<0.03	total 0.342
50 SL	14/28	BR	<0.03/<0.03	0.02
50 SL	14/28	BR	<0.03/<0.03	0.04

[1] IND = India; ET = Egypt; T = Thailand; GCA = Guatemala; BR = Brazil

Table 68. Residues of cyfluthrin in olives (Internal Residue Report, Bayer AG, Leverkusen, FRG)

Formulation	Days after last application	Country[1]	Residue levels (mg/kg) at harvest or at days after last application	Dosage a.i. (kg/ha)
50 SL	40	I	0.04	0.068

[1] I = Italy

According to N. P. Agnihotri [142] this substance is considered to be more effective at higher temperatures and more photostable than e.g. fenvalerate and has in addition a wider range of application because it has aphicidal as well as acaricidal activity.

Further residue data in cotton shown in Table 67 cover the residue situation of cyfluthrin.

Only one report on residues of cyfluthrin in olives is available from Italy (Table 68).

2.2.4.2 Hops

Data on residues of cyfluthrin in hops are presented in Table 69.

2.2.4.3 Alfalfa

Residues of fenvalerate and cypermethrin on Alfalfa can be taken from Table 70. [52] contains additional information about effect of growth dilution upon the residue level and about the influence of temperatures upon the residue situation.

Concerning the results reported under [143] it should be added that a MRL value for cypermethrin on alfalfa has not yet been established. The residue levels [143] decline to approx. 20% of the initial residues within 17 days after treatment.

Residue levels of cyfluthrin in alfalfa after one spray treatment with the 2 EC formulation are presented in Table 71.

Table 69. Residues of cyfluthrin in hops (Internal Residue Reports, Bayer AG, Leverkusen, FRG)

Formulation	Days after last application	Country[1]	Residue levels (mg/kg) at harvest or at days after last application	Dosage a.i. (kg/ha)
25 EC	65	B	<0.4	total 0.105
50 EC	0/3/5/7/10	D	12.7/ 10.2/ 7.1/ 9.7/ 6.0	total 0.406
50 EC	0/3/5/7/10	D	13.3/ 12.3/ 6.2/ 5.4/ 4.2	total 0.406
50 EC	0/3/5/7/10	D	10.3/ 8.7/ 8.0/ 5.6/ 5.5	total 0.406
50 EC	0/3/5/7/10	D	7.5/ 5.9/ 6.3/ 3.9/ 2.3	total 0.333
25 EC	29/51	B	4.6/ 2.2	1 × 0.047
50 EC	0/5/7/10/14	D	24.9/ 16.6/ 4.6/ 6.9/ 8.6	total 0.55
50 EC	0/5/7/10/14	D	24.7/ 24.5/ 16.0/ 17.5	total 0.825
50 EC	0/5/7/10/14	D	26.0/ 11.8/ 15.2/ 18.7/ 6.4	total 0.825
50 EC	0/5/7/10/14	D	16.5/ 19.1/ 11.4/ 12.6/ 12.5	total 0.825
500 EC	0/21/35/42/49	D	2.3/<1.0/<1.0/<1.0/<1.0	
500 EC	0/21/35/42/49	D	24.1/<1.0/<1.0/<1.0/<1.0	
500 EC	0/21/35/42/49	D	28.8/ 1.3/ 1.2/<1.0/<1.0	
500 EC	0/21/35/42/49	D	5.2/<1.0/<1.0/<1.0/<1.0	
50 EC	0/5/7/10/14	D	1.4/<1.0/<1.0/ 2.1/ 2.5	1 × 50/75/ 125/200/ 200 g/ha
50 EC	0/5/7/10/14	D	20.4/ 14.4/ 13.4/ 7.8/ 6.6	1 × 125/
50 EC	0/5/7/10/14	D	28.9/ 8.0/ 6.7/ 7.4/ 4.8	150/175/200/
50 EC	0/5/7/10/14	D	27.1/ 7.2/ 7.2/ 6.5/ 2.3	250 g/ha
50 EC	0/3/7/13	CS	7.0/ 5.5/ 3.0/ 2.1	0.05
	0/5/7/10/14		1.0/ 1.1/ 1.2/ 1.6/ 1.1	
	0/5/7/10/14		2.6/ 0.5/ 1.2/ 1.1/ 1.8	
	0/5/7		1.4/ 1.2/ 1.0	

[1] B = Belgium; D = Federal Republik of Germany; CS = Czechoslovakia

2.2.4.4 *Tobacco*

Table 72 covers residues of pyrethroids on/in tobacco. The following additional information is to be given concerning the results presented in this Table under [59]. The flue-curing (drying) of the green tobacco resulted in a reduction of the residues of about 60% in both test years (1980 and 1981). Permethrin was more persistent on green and on flue-cured tobacco than fluvalinate.

W. B. Tappan et al. [144] reported that fenvalerate and permethrin have only limited insecticidal properties on tobacco and result in high residues on the cured leaves (Table 73).

Tank mixes with insecticidal organic phosphates increased the spectrum of control. As was to be expected, the pyrethroid residues on tobacco were more persistent than those of the organic phosphates.

Some additional data covering residue levels of cyfluthrin in in tobacco leaves are shown in Table 74.

Table 70. Residues of pyrethroids in alfalfa

Compound	Formulation	Days after last application	Country[1]	Residue levels (mg/kg) at harvest or at days after last application	Dosage a.i. (kg/ha)	Remarks (Ref.)
Fenvalerate	30 EC	0/10	CAN	27/~13.5[2]	1 × 0.14—0.15	52)
Cypermethrin	5 EC	0/1/2/4/8/10/14/17	SAUDI-ARABIA	10.5/8.7/8.4/6.5/3.8/3.5/2.9/2.0	"0.2%"	143)

[1] CAN = Canada; [2] Mean values of 3 experiments

Table 71. Residues of cyfluthrin in alfalfa (Internal Residue Reports, Bayer AG, Leverkusen, FRG)

Formulation	Days after last application	Country[1]	Residue levels (mg/kg) at harvest or at days after last application	Dosage
2 EC	3/8/24/48/72/96 144/192/216/240 264/312/336	D D D	3.63/3.17/3.39/3.60/5.09/3.79 3.39/1.36/1.33/1.72 1.65/1.95/1.65	1 × 0.056
2 EC	0/3/8/24/48/72/96 120/168/192	D D	5.22/4.79/5.22/2.91/2.27/1.22/2.38 1.80/1.74/1.97	1 × 0.056
2 EC	0/3/8/24/48/72 120/144	D D	1.29/1.57/1.18/0.86/1.19/0.91 0.96/0.88	1 × 0.056

[1] D = Federal Republik of Germany

Table 72. Residues of pyrethroids in tobacco

Compound	Formulation	Days after last application	Country[1]	Residue levels (mg/kg) at harvest or at days after last application	Dosage a.i. (kg/ha)	Remarks (Ref.)
Fluvalinate	n.g.	0/12	USA	3.3/0.3	0.11	[59]
				in green tobacco		
		0/12		7.2/2.7	0.22*)	
		0/16		1.7/0.4	0.11	
				in green tobacco		
		0/16		3.0/0.5	0.22**)	
	n.g.	0/16		0.9/0.3	0.056	
				in green tobacco		
		0/16		1.3/0.6	0.11***)	
Permethrin	n.g.	0/12		9.6/7.0	0.22*)	
		0/16		3.5/3.2 in green tobacco	0.11**)	
		0/16		3.8/3.5	0.22**)	

n.g. = not given; [a] Experiments in 1980; [b] Experiments in 1981; [c] Experiments in 1981 with a new fluvalinate formulation

Table 73. Chemical residues from sprays of organophosphates and synthetic pyrethroids applied alone or in tank mixes on green and cured flue-cured tobacco, Live Oak, Fla., 1979; according to W. B. Tappan et al. [144]

Treatment	Harvest no.	Rate (kg of AI/ha)	Bo. of applications	Days from last application	Residues on tobacco (ppm)[a, b, c]	
					Green	Cured
Acephate	1	0.53	4	5	ND	ND
	2	0.54	5	6	ND	ND
+	3	0.50	6	13	ND	ND
	4	0.50	6	20	ND	ND
Permethrin	1	0.06	4	5	ND	7.7
	2	0.06	5	6	4.8	—
	3	0.06	6	13	7.6	7.0
	4	0.06	6	20	6.4	9.4
Encapsulated parathion-methyl	1	0.65	4	5	ND	ND
	2	0.63	5	6	ND	ND
+	3	0.63	6	13	ND	ND
	4	0.63	6	20	ND	ND
Permethrin	1	0.06	4	5	3.6	16.2
	2	0.06	5	6	2.6	—
	3	0.06	6	13	2.3	8.9
	4	0.06	6	20	3.1	7.3
Acephate	1	0.63	4	5	ND	ND
	2	0.61	5	6	ND	ND
+	3	0.60	6	13	ND	ND
	4	0.60	6	20	ND	ND
Fenvalerate	1	0.06	4	5	ND	3.0
	2	0.06	5	6	ND	6.5
	3	0.06	6	13	4.5	4.3
	4	0.06	6	20	>1.0	3.6
Permethrin	1	0.18	4	5	12.4	50.7
	2	0.18	5	6	9.3	—
	3	0.17	6	13	13.2	28.8
	4	0.17	6	20	6.5	22.2

Fenvalerate	1	0.14	4	5	3.9	16.5
	2	0.13	5	6	>1.0	20.1
	3	0.13	6	13	>1.0	12.3
	4	0.13	6	20	ND	12.4
Acephate	1	0.86	4	5	ND	ND
	2	0.87	5	6	ND	ND
	3	0.84	6	13	ND	ND
	4	0.84	6	20	ND	ND
Encapsulated parathion-methyl	1	0.67	4	5	ND	ND
	2	0.65	5	6	ND	ND
	3	0.63	6	13	ND	ND
	4	0.63	6	20	ND	ND

[a] Residue figures for permethrin are cis- and trans-isomers combined;
[b] Means of duplicate analyses;
[c] ND, None detected with method sensitivity of 0.1 ppm.
Formulation types were not given. The rates of application were 0.13 kg a.i./ha for fenvalerate and 0.18 kg a.i./ha for permethrin respectively.

Table 74. Residues of cyfluthrin in tobacco leaves (Internal Residue Reports, Bayer AG, Leverkusen, FRG)

Formulation	Days after last application	Country[1]	Residue levels (mg/kg at harvest or at days after last application	Dosage a.i. (kg/ha)
	15	E	0.6	1 × 0.025
50 EW	3	E	1.4	1 × 0.025
50 EC	3	E	0.6	1 × 0.025

[1] E = Spain

2.2.4.5 Tea

Residues of pyrethroids on tea can be taken from Tables 75 and 76. Concerning the information given under [91], a total of 21 samples of imported tea were analysed. The Danish MRL is 20 mg/kg as given by FAO/WHO. The values reported under [145] are based on 99 analysed samples of black tea. These samples were taken at the importer and/or the wholesale dealer. Only the trans-permethrin was detected in the one sample which contained permethrin residues.

According to [146] fluvalinate is able to replace the highly persistent and bioaccumulating chlorinated hydrocarbon insecticides in tea cultivation. The persistence of fluvalinate is markedly lower.

A MRL for fluvalinate on tea was not yet established, yet a value in the order of magnitude of about 20 mg/kg as it exists already for cypermethrin and permethrin in this crop, is discussed by Gopal et al. [146].

Only 0.2% of the active ingredient were transferred into the liquor when preparing tea by adding boiling water to spiked tea leaves (20 mg/kg). Consequently no hazards result from residues. The safety intervals are covered by the production process for the tea because tea is principically picked seven days after the spray application. In addition the tea is stored after manufacture. Based on the data reported under [146] the calculated half-lives of fluvalinate residues on tea leaves are between 2 and 5 days.

Table 75. Residues of pyrethroids in tea

Compound	Formulation	Days after last application	Country[1]	Residue levels (mg/kg) at harvest or at days after last application	Dosage a.i. (kg/ha)	Remarks (Ref.)
Permethrin	unknown[2]	unknown[2]	DK	3.71	—	[91]
Permethrin	unknown[2]	unknown[2]	DK	3.71[3]	—	[145]

[1] DK = Denmark; [2] results from official Danish market control; [3] only trans-permethrin was detected

Table 76a. Wet weather residues of fluvalinate in processed tea; according to M. Gopal et al [146]

Days	Manufactured				Sundry			
	Residue in ppm				Residue in ppm			
	R_1	R_2	R_3	Avg.	R_1	R_2	R_3	Avg.
(A)	DILUTION 1:2000 (normal dose: 0.0125% a.i.)							
0	4.17	1.90	3.07	3.04	2.41	3.10	3.79	3.10
1	3.00	2.01	2.20	2.40	3.02	2.55	2.12	2.56
3	2.10	1.97	1.84	1.97	2.20	2.11	1.90	2.07
7	0.91	0.66	0.77	0.78	0.83	1.08	1.36	1.09
15	0.57	0.20	0.35	0.34	0.19	0.35	0.52	0.35
(B)	DILUTION 1:1000							
0	4.73	4.28	4.52	4.51	4.07	3.94	4.00	4.01
1	4.29	3.36	3.79	3.81	3.76	3.88	3.85	3.82
3	3.46	3.45	3.51	3.46	1.66	0.71	1.19	1.18
7	0.69	0.48	ND	0.39	0.16	0.13	0.15	0.15
15	0.30	ND	0.25	0.18	ND	ND	ND	—

[a] 3 replicate experiments

Table 76b. Dry weather residues of fluvalinate in processed tea

Days	Manufactured				Sundry			
	Residue in ppm				Residue in ppm			
	R_1	R_2	R_3	Avg.	R_1	R_2	R_3	Avg.
(C)	DILUTION 1:2000 (normal dose: 0.0125% a.i.)							
0	11.22	11.63	12.05	11.64	13.18	12.72	12.25	12.72
1	7.58	6.64	5.69	6.64	6.98	4.45	9.51	6.98
3	4.15	4.51	6.13	4.29	7.77	5.96	6.87	6.87
7	1.88	1.51	1.89	1.76	2.37	1.64	2.02	2.01
15	0.24	0.21	0.26	0.24	0.23	ND	0.12	0.12
(D)	DILUTION 1:1000							
0	10.39	9.69	9.00	9.69	20.06	20.45	19.68	20.06
1	12.69	6.62	9.65	9.65	14.02	12.82	12.52	13.12
3	4.67	4.89	4.56	4.71	7.32	7.02	6.98	7.12
7	2.34	2.89	3.46	2.89	2.04	3.43	2.70	2.74
15	0.69	0.54	0.39	0.54	0.47	0.65	0.29	0.47

[3] replicate experiments;
ND = not detectable

2.2.4.6 Soil

Pyrethroids may also be used as soil insecticides. Respective soil after the use of a pyrethrin product which contained 6.45% of pyrethrins and 25.8% of piperonyl butoxide as synergist, is demonstrated under [68]. A good residual effectiveness in the soil resulted. The residues declined from an initial level of about 3 mg/kg to 1 mg/kg within 28 days after the treatment.

Table 77. Residual pyrethrins in the soil (Finland, 1977)

Sampling	Days from application	Moisture %	Residual pyrethrins ppm, dry weight
7. 6. 77	1	11.6	3.1
10. 6. 77	4	14.6	2.7
14. 6. 77	8	5.0	2.4
21. 6. 77	15	8.8	1.2
28. 6. 77	22	10.3	1.3
4. 7. 77	28	18.5	0.7
12. 7. 77	36	7.3	0.3
19. 7. 77	43	15.5	0.2
26. 7. 77	50	17.6	0.1
9. 8. 77	64	11.5	0.1
30. 8. 77	85	13.8	0.1
12. 10. 77	128	17.5	0.1

Table 77 is according to H. Siltanen et al. [68];
The product used consisted of 6.45% pyrethrins and 25.8% piperonyl butoxide (0.1% a.i.; 2 l/m^2).
The top 3 cm layer was sampled for analysis. Soil type: sandy soil. Average temperature from May to September 1977: 13.4 °C, rainfall: 305 mm; hours of sunshine: 144

According to [94] in Table 78, a granular formulation of permethrin was incorporated to a depth of 3 cm. The residues persisted for about 28 days. The substance was gradually decomposed during the rest of the season. A translocation of active ingredient into edible portions of cucumbers and head lettuce was not observed. On the other hand permethrin residues were detectable in the root system of the two vegetables.

Residues of permethrin, cypermethrin, fenvalerate and deltamethrin in soil are summarized in Table 79 as results of an Indian study [157].

It can be taken from the data that permethrin, cypermethrin and fenvalerate persisted for 30–40 days. Deltamethrin residues were no longer detectable after 7 days after application of 12.5 g a.i./ha and after 15 days after application of 25 g a.i./ha respectively. It is reported in this publication that no leaching of residues could be observed at the normal rate of application. Only traces of the active ingredients, if any, leached into deeper soil layers at the higher rates of application. After 30 days, however, these traces were no more detectable.

The following information concerning the residue situation of pyrethroids in soil was taken from literature and could not be summarized in tabular form. Thus T. R. Roberts and M. E. Standen [147] reported that wheat plants absorbed radioactively labeled cypermethrin from treated soil only to a minor extent. This soil contained radiolabeled bound residues. Only an amount of 0.14 to 0.58% of the radioactivity in the soil was detected in the wheat.

After an extensive search of the literature concerning the residue situation of pyrethroids in soil the conclusion of W. Ebing [148] was: there is no risk of an impact the soil or on the consumer.

J. Lhoste and M. L'Hotellier [149] found that deltamethrin remains in the upper 2.5 cm layer of the soil. This applies to different soil types. According to the authors the strong adsorption of deltamethrin to the soil particles is accompanied by an addi-

Table 78. Residues of pyrethroids in the soil

Compound	Formulation	Days after last application	Country[1]	Residue levels (mg/kg) at harvest or at days after last application	Dosage a.i. (kg/ha)	Remarks (Ref.)
Permethrin	Granul.	0/7/14/28/63/113	CAN	13.6/13.4/14.2/13.8/6.8/7.9	1.12	94)
		0/7/28/63/113		6.9/7.6/7.0/6.2/5.1/4.2	0.56	

[1] CAN = Canada

Table 79. Residues of permethrin, cypermethrin, fenvalerate and deltamethrin in soil (India); according to N. P. Agnihotri et al. 157)

Insecticide	Dose (g a.i./ha)	Depth (cm)	Residues in soil (ppm)						Half-life (days)
			1 hr	3 days	7 days	15 days	30 days	45 days	
Permethrin	125	0.0—7.5	0.107	0.086	0.049	0.016	0.004	ND	5.8 +/— 0.5
		7.5—15.0	—	—	—	—	—	—	
	250	0.0—7.5	0.212	0.149	0.085	0.042	0.005	ND	
		7.5—15.0	—	0.006	0.004	0.006	ND	ND	
Cypermethrin	75	0.0—7.5	0.068	0.023	0.015	0.003	ND	ND	3.4 +/— 0.3
		7.5—15.0	—	—	—	—	—	—	
	150	0.0—7.5	0.111	0.050	0.034	0.007	ND	ND	
		7.5—15.0	—	0.009	0.007	0.003	ND	ND	
Fenvalerate	100	0.0—7.5	0.100	0.064	0.033	0.006	0.002	ND	5.0 +/— 0.1
		7.5—15.0	—	—	—	—	—	—	
	200	0.0—7.5	0.165	0.110	0.060	0.014	0.003	ND	
		7.5—15.0	—	ND	ND	ND	ND	ND	
Deltamethrin	12.5	0.0—7.5	0.01	0.005	ND	ND			3.0 +/— 0.5
		7.5—15.0	—	ND	ND				
	25.0	0.0—7.5	0.025	0.01	0.004	ND			
		7.5—15.0	—	ND	ND				

ND = not detectable

tional fast biodegradation of the substance. Therefore they see no risk of accumulation as would be expected to occur with most of the chlorinated hydrocarbon insecticides.

C. R. Harris et al. [150] concluded that soil type, soil moisture and temperature have only a minimal influence upon the toxicity of cypermethrin, fenvalerate, fenpropanate and permethrin. The persistence of these four pyrethroids in soil was superior or equal to that of the carbamate soil insecticide carbofuran.

N. S. Talekar et al. [151] reported about the persistence of fenvalerate in subtropical soil after repeated biweekly applications over two years. No carryover from year to year was determined by the authors. An increased degradation of the substance occurred from May to October, as during this period higher temperatures and heavier rainfalls were recorded. Leaching of the fenvalerate below the 0–15 cm layer was not observed. As an example, a soil residue of 0.008 mg/kg resulted 6 months after 2 applications per week at a rate of 0.2 kg a.i./ha.

B. D. Hill [152] investigated the soil residue situation in an indoor incubation study up to 52 weeks after application of 17.5 g of deltamethrin/ha (2.5% EC formulation). The deltamethrin residues declined at a half-life of 6.9 weeks with 5–7% of the applied deltamethrin remaining after 52 weeks.

S. Smith and G. H. Willis [153] determined that the persistence of parathion-methyl in soil in the presence of flucytrinate increased 1.5 to 2-fold within an observation period of 3–6 weeks. The persistence of permethrin in the presence of methyl-parathion was reduced by 25–40% within 6 weeks. Such interactions of two insecticides show that it is necessary to consider more than just the degree of insect pest control when applying combinations of active ingredients. Under these aspects changes in the environmental behaviour of the compound must also be taken into consideration.

C. H. Schaefer et al. [154] reported about residues of fenvalerate in pasture soil after multiple aerial applications. The highest residue level found in this connection was at 0.3 mg/kg. The soil residues declined after each application but they were still detectable 56 days after the third application.

N. P. Agnihotri and H. K. Jain [155] analysed for soil residues under field conditions after application of flucythrinate and fluvalinate (50 and 100 g a.i./ha). Fluvalinate persisted for 30 days at both rates of application while flucythrinate persisted for 30 days only at the low rate. At the high rate flucythrinate persisted for 40 days. The half-lives were indicated to be as follows: fluvalinate 6.8–8.0 days and flucythrinate 9.4–11.9 days.

M. Bahadir et al. [156] investigated the persistence of cypermethrin, deltamethrin, fenvalerate and permethrin in a standardized soil test under laboratory conditions with and without formulation of the active ingredients. Degradation of unformulated active ingredients occurred with the following half-lives: 21 days with cypermethrin, 88 days with fenvalerate, 110 days with deltamethrin and 126 days with permethrin. Formulation with inorganic adsorbents such as silica gel had no influence upon the rates of degradation. The persistence of cypermethrin, however, increased considerably when being formulated with sawdust of different origin (in case of oak 110 days). The persistence increased in case of all active ingredients when using oak dust formulations ($t_{1/2}$ = 110—475 days).

2.2.5 Products Originating from Animals, Including Tissues (Meat, Innards, Fat), Milk and Eggs

2.2.5.1 Cows

Concerning the residues presented in Table 80 in [58] it can be said that a withholding period for milk and tissues is not required based on these findings.

In the same Table residues in milk are shown under [165] which resulted after treatment of the animal with a relatively high dose of 10 mg of deltamethrin per kg of body weight. About 90% of these residues were found in the milk fat.

Only a very low degree of accumulation of deltamethrin was observed in the edible tissues. About 40% of the administered radioactivity was eliminated with the faeces within 24 hours after the last administration. Unchanged deltamethrin was the main component constituting about 80% of the radioactivity. The residues in different tissues were usually very low (in the range of 0.1 μg/g). They were only higher in liver, kidney, udder, abdominal and subcutaneous fats (between 0.28 and 3.19 μg/g).

Residues of deltamethrin in whole milk after oral or dermal treatment or after treatment of the stable walls are reported under [76] in Table 80. The "half-life" in whole milk was 0.8 days after intrarumenal administration and 4.5 days after dermal application. The maximum residue limits were reached after 2.5 days. Calculating these figures for rendered butter fat, the residue levels are correspondingly higher, namely 0.92 mg/kg (attained after 1 day) after oral administration or 0.10 mg/kg (attained after 2 days) after dermal application, respectively.

In muscular tissue (leg) the residues were 0.008 mg/kg (oral route) 2 days after treatment and in omental fat they were 0.088 mg/kg (oral route) 2 days after treatment. Also in this study [76] the residues were identified as predominantly unchanged deltamethrin.

The results reported under [164] show that even at the high dose of 10 μg/g of feed only low residues of cypermethrin occur in cow milk (0.03 mg/kg). Ingestion and elimination of applied radioactivity were in balance in the milk at about day 4 after the start of soding (see Fig. 8). The residues in the milk consisted of unchanged cypermethrin.

In the course of this study [164] the respective muscular substance was also analysed for cypermethrin residues. The radioactivity measured in the muscle was however, so low that the chemical nature was not investigated.

The results listed under [172] in Table 80 are based on a treatment with cypermethrin by means of impregnated ear tags. These tags contained 7% cypermethrin, 3.5% piperonyl butoxide and also 5% chloropyrifos. The weight of one tag was 9.6 g.

Neither cypermethrin nor piperonyl butoxide residues were detectable (<0.04 and <0.05 mg/kg resp.). The residue situation in fat samples of the respective animals (perianal fat, diaphragmatic fat, omental fat, renal fat) is the same. The EPA tolerance for cypermethrin in cattle fat is 0.05 mg/kg and was therefore not reached in this study. Also in muscle, heart, liver and kidney no residues of cypermethrin and piperonyl butoxide were detectable (<0.04 and <0.05 mg/kg resp.). The following EPA tolerances were established for these materials: meat 2.0 mg/kg and meat by-products: 0.1 mg/kg.

Also the residue data presented in Table 80 under [163] resulted after cypermethrin treatment by means of impregnated ear tags. The tags consisted of molded PVC and had a diameter of 60 mm. The active ingredient content in the tag was 806 mg. Milk

Table 80. Residues of pyrethroids in animal products (cows)

Compound	Formulation	Days after last application	Country[1]	Residue levels (mg/kg) at harvest or at days after last application	Dosage a.i. (kg/ha)	Remarks (Ref.)
Flumethrin	EC	6 hrs/18 hrs/1/2	R.S.A.	<0.01 in milk (fat phase	spraying with a 75 mg/kg solution in 2-weeks intervals	58)
		1/3	AUS	<0.05 in muscle, fat, liver and kidney of steers	50/100 and 200 mg/kg (10 liters per animal)	58)
Deltamethrin	^{14}C-Exp.	1	CAN	0.10—0.14 in milk	10 mg/kg of body weight at 3 consecutive days, oral admin.	165)
Deltamethrin	—	1/3 2.5 1	F	0.045/0.0038 in whole milk 0.0057 in whole milk 0.00125 in whole milk	0.27 g oral[a] 0.21 g dermal 0.0075 g/m^2, wall treatment	76)
Cypermethrin	^{14}C-Exp.	4 days after initial dose	UK	0.3 in milk 0.01 in milk <0.002 in milk	10 µg/g of diet[a] 5 µg/g of diet 0.2 µg/g of diet	164)
Cypermethrin	impregnated ear tag	14 weeks post treatment 10, 12 and 14 weeks post treatment	USA	<0.04 in muscle, heart, liver and kidney <0.04 in fatty tissues	see text	172)
Cypermethrin	impregnated ear tags	21-day period	CAN	residues in milk only in 15% of the samples (mean value: 6.4 ± 2.1 µg/kg of butterfat); kin 85% of the samples: <4 µg/kg fat	see text	163)

Resmethrin	^{14}C-Exp.	12—48 h after treatment	USA	0.107 trans isomer, TAC, milk[b] 0.055 trans isomer, TAL, milk 0.092 cis isomer, CAC, milk 0.190 cis isomer, CAL, milk (all the values given as % of administered radiocarbon)	10 mg/kg of body weight as single oral doses	171)
Fenvalerate	—	6 h/3/4 (time from spray application to milk sampling	CAN	0.0002/0.00046/0.0002 (milk)[c]	100 mg a.i./animal[d]	162)
		6 h		0.00105 (first treatment), milk[c]	500 mg a.i./animal[e]	
		6 h		0.00419 (second treatment), milk		
		6 h		0.00206 (third treatment), milk		
		3		0.00092 (first treatment), milk		
		3		0.00169 (second treatment), milk		
		3		0.00091 (third treatment), milk		
		21		0.0002 (third treatment), milk		
Permethrin	—	5	D	0.012	2 mg/kg of body weight, dermal appl.	169)
Permethrin	—	0/4/7	D	0.03/0.015/0.005	2 mg/kg of body weight, dermal appl.	170)
		0/4/7		0.5 /0.07 /0.015	10 mg/kg of body weight, dermal appl.	
		0/4/7		0.4 /0.17 /0.03	20 mg/kg of body weight, dermal appl.	

[1] R.S.A. = Republic of South Africa; AUS = Australia; CAN = Canada; F = France; UK = United Kingdom; D = Federal Republic of Germany.
[a] for further details see text;
[b] TAC = acid ^{14}C-(1RS)-trans-resmethrin; TAL = alcohol ^{14}C-(1RS)-trans-resmethrin, CAC = acid ^{14}C-(1RS)-trans-resmethrin, CAC = acid ^{14}C-(1RS)-cis-resmethrin;
[c] CAL = alcohol ^{14}C-(1RS)-cis-resmethrin, for further details see text;
sum of the diastereoisomers, mean values from two cows;
[d] six consecutive topical treatments over a 26-day period;
[e] three consecutive topical treatments over a 59-day period

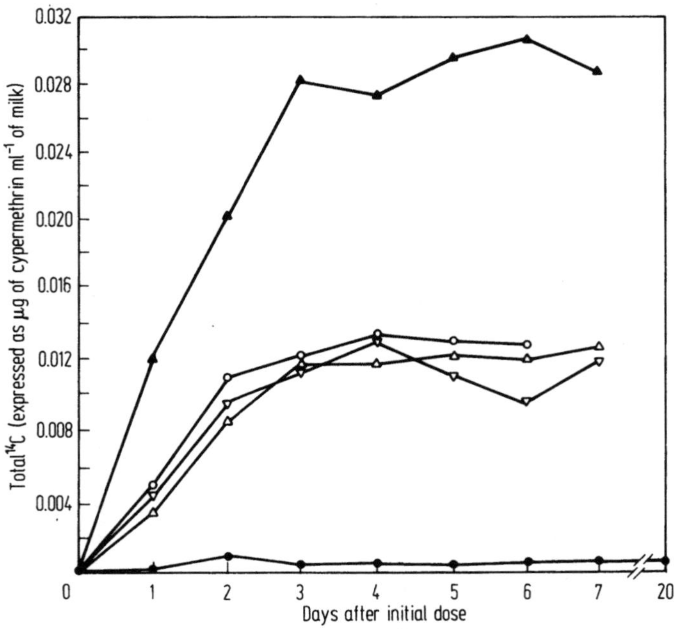

Fig. 8. Radioactivity in the combined morning and afternoon milks of cows treated daily with various dietary concentrations of benzyl- or cyclopropyl-labelled [^{14}C]cypermethrin: (●) cows 1 and 2, 0.2 µg [benzyl-^{14}C]cypermethrin g^{-1}; (○) and (□) cows 3 and 4, 5 µg [cyclopropyl-^{14}C]cypermethrin g^{-1}; (△) cow 5, 5 µg [benzyl-^{14}C]cypermethrin g^{-1}; (■) cow 6, 10 µg [benzyl-^{14}C]cypermethrin g^{-1}. (according to A. Croucher et al. [164])

samples were taken from a total of 10 cows over a 21-day period. Cypermethrin residues were detectable in the milk of 5 of these animals; the milk of the other animals contained residues below the limit of determination of 4 µg/kg, calculated on the basis of milk fat. A total of 71 samples were analysed of which 60 samples or about 85% contained residues below 4 µg/kg. In 11 samples, residues were between 4 and 9.6 µg/kg, again calculated on the basis of milk fat. The mean value was 6.4 ± 2.1 µg/kg of fat. The Food and Drug Act and Regulations (1979) define negligible residues as being <0.1 mg/kg in food commodities. The highest residue levels of cypermethrin in milk fat determinated in this study are at about 1/10 of the negligible residue level.

The residue data presented in the same Table under [171] are from a metabolism study with ^{14}C-labeled cis- and trans-resmethrin. In this case an oral dose of 10 mg/kg of body weight was administered to milk cows. The elimination of radioactivity with the milk was low, amounting to 0.055 to 0.190% of the administered radioactivity. The radioactivity was mainly eliminated with the faeces and the urine. Higher residues of >1 mg/kg occurred in this study only in liver and kidney. In case of TAC and TAL (for explanation of these terms see Table 80) each of these residues was higher in the kidney than in the liver. In case of CAC and CAL, in turn, the ^{14}C-contents determined in the liver were higher than those in the kidney by the factor of 2. Owing to the rapid metabolisation and elimination with the faeces and the urine, resmethrin can be classified as being well suitable for stable treatments [171].

Table 81. Residues of pyrethroids in sheep and lamb tissues

Compound	Formulation	Days after last application	Country[1]	Residue levels (mg/kg at harvest or at days after last application	Dosage a.i. (kg/ha)	Remarks (Ref.)
Deltamethrin	—	2/7	F	0.018/ 0.016 (muscle) 0.011/ 0.007 (kidney) 0.001/<0.001 (liver)	10 mg/kg oral dose	76)
		7/14/21		<0.001/ 0.001/<0.001 (muscle)	2 mg/kg dermal appl. ("pour on")	
Fenvalerate	—	—	USA	0.22 (muscle)[a] 0.16 (kidney) 0.1 (liver) 4.0 (fat)	45 mg/kg in ration during a 10-day period	174)

[1] F = France;
[a] mean values of two lambs, the animals were fed with 45 mg/kg a.i. in ration and consumed 4.5 kg of diet within 10 days, equivalent to 0.45 kg of diet per day corresponding to 20.25 mg of a.i./day

Table 82 a. Residues of pyrethroids in chickens

Compound	Formulation	Days after last application	Country[1]	Residue levels (mg/kg) at harvest or at days after last application	Dosage a.i. (kg/ha)	Remarks (Ref.)
Cypermethrin	^{14}C-Exp.	8 days after the start of dosing 4 h after the last of 14 daily oral doses	UK	0.05 µd/g of whole egg[a] 0.1 µg/g of albumen 0.15 µg/g of yolk 0.012 µg/g breast muscle 0.022 µg/g leg muscle 0.08 µg/g fat (peritoneal) 0.08 µg/g fat (subcutaneous) 0.37 µg/g liver	1.52 mg a.i. per day, oral, for 14 days	160)
Deltamethrin	^{14}C-Exp.	1/2/3/4/5 after the last dose	CAN	0.155/0.193/0.026/0.005/n.d. µg/g albumen[b] 0.576/0.828/0.700/0.658/0.586 µg/g yolk[b]	7.5 mg a.i. on each of 3	158)

[1] UK = United Kingdom; CAN = Canada;
[a] For further details see text;
[b] Sum of ^{14}C-gem-dimethyl- and benzyl-^{14}C-deltamethrin, see text

Table 82b. Amount of Deltamethrin Equivalent (μg/g) in Tissues of Chickens Given Orally 7,5 mg of ^{14}C-Labelled Insecticide Daily on Each of Three Consecutive Days[a]

Tissues	Gem-dimethyl									
	6 h		18 h				48 h		120 h	
	15[b]	47	2	4	6	8	1	5	7	17
Liver	0.27	0.49	1.89	2.43	3.95	1.89	0.52	0.44	0.58	0.12
Kidney	0.43	0.79	4.59	3.75	6.85	3.36	0.86	1.05	0.25	0.26
Heart	0.02	0.02	0.61	0.31	0.43	0.21	0.05	0.07	0.09	0.04
sq. fat	lost	0.14	0.47	0.12	0.26	0.39	0.17	0.12	0.04	0.19
Abdominal fat	lost	0.15	0.66	0.27	0.41	0.25	0.18	0.07	0.08	0.13
Breast muscle	0.03	0.02	0.14	0.09	0.21	0.08	0.18	0.09	0.01	c
Leg muscle	0.02	0.02	0.10	0.15	0.14	0.09	0.03	0.03	0.01	c
Ovarian yolk	0.31	0.76	3.31	1.35	1.21	2.05	0.67	0.69	c	0.06
Gizzard	0.05	0.02	0.27	0.35	0.61	0.29	0.06	0.08	0.04	0.01
Blood	0.07	0.11	0.79	1.01	1.01	0.41	0.13	0.15	0.05	0.08

Tissues	Benzyl									
	6 h		18 h				48 h		120 h	
	35	37	14	16	18	20	21	29	19	23
Liver	0.35	0.19	0.63	0.36	0.89	0.97	0.23	0.19	0.06	0.09
Kidney	0.72	0.27	1.29	1.19	2.11	2.52	0.25	0.52	0.11	c
Heart	0.09	0.02	0.09	0.04	0.07	0.08	0.04	0.03	c	c
sq. fat	1.95	0.09	0.16	0.08	0.13	0.17	0.37	0.11	0.12	0.11
Abdominal fat	c	0.12	0.17	0.11	0.09	0.33	0.25	0.13	0.13	0.12
Breast muscle	0.04	0.02	0.03	c	c	0.15	0.03	0.03	c	c
Leg muscle	0.12	0.02	0.01	0.09	0.02	0.05	0.03	0.03	c	c
Ovarian yolk	0.45	0.32	0.54	0.79	0.78	0.53	0.43	0.16	0.02	0.08
Gizzard	1.18	0.03	0.12	0.03	0.24	0.23	0.07	0.07	c	0.04
Blood	0.12	0.04	0.15	0.18	0.25	0.34	0.14	0.06	0.02	0.02

[a] Value is an average of two or three determinations at the indicated time after the last dose;
[b] Refers to bird number;
[c] Below 5 ppb level;
Table 82 (cont.) is according to M. H. Akhtar et al [158]. See text for further details

As reported in Table 80 [167], none of the diastereoisomers of fenvalerate was detectable in milk 6 hours after topical application of 100 mg a.i./animal. In this dose group the fenvalerate residues averaged 0.46 μg/l of milk 3 days after application and only less than 0.2 μg/l for the total of both diastereoisomers after 4 days. In case of the higher dose of 500 mg a.i./animal the highest residues were measured 6 hours after treatment. Within 21 days these residues reached the level of less than 0.2 μg/l. Only 0.03 to 0.06% of the applied amount of active ingredient appeared in the milk as unchanged parent compound.

In this study the application was made by hand all over the animal. Of interest is the finding reported in this study [167] that there are no differences in the content of the two diastereoisomers in the milk.

In the study cited under [169] in Table 80 permethrin was applied as ectoparasiticide. The active ingredient was applied dermally to milk cows at an amount of 2.0 mg/kg of body weight. In this study about 0.2% of the applied amount of permethrin were eliminated with the milk.

After dermal application of 2 mg of permethrin per kg of body weight the residues in milk were 0.01 mg/kg over a sampling period of 5.5 days. In case of the higher dose groups of 10 and/or 20 mg/kg of body weight the residues were respectively higher (see Table 80 under [170]). On the other hand no residues, i.e. in this case <0.005 mg/kg were detectable in milk after spraying a 1% solution of permethrin (35 ml/m^2) in a stable.

P. C. Wszolek et al. [168] reported about the elimination of fenvalerate in milk after treatment of 2 cows each with 5 and 15 mg/kg respectively. The active ingredient was administered to the cows for 4 days in 22.7 kg of feed/day. Thus, the total amount of active ingredient administered was 0.454 g in case of the 5 mg/kg dose and 1.362 g in case of the 15 mg/kg dose, respectively. The elimination in milk was 0.44% in case of the lower dose and 0.64% of the administered amount in case of the higher dose respectively. Incidentally, about 25% of the respective dose was eliminated in this study with the faeces. The analytical determinations covered only the unchanged parent compound while no analyses for metabolites were made.

2.2.5.2 Sheep/Lamb

According to [76] in Table 81 no residue problems arise in veterinary use if sheep are treated for the control of ectoparasites with 10 mg of deltamethrin per kg orally or 2 mg/kg dermally, respectively.

The results of a study concerning the accumulation in lamb tisses after administration of fenvalerate with the feed can be taken from Table 81 under [174]. In this study it turned out that the active ingredient accumulates in the fat of the animals to quite an extent.

2.2.5.3 Chickens

Residue data of pyrethroids in chickens or laying hens resp. are summarised in Table 82. The values referenced in Table 82 under [160] are from a ^{14}C-phenoxy cypermethrin study in which the active ingredient was administered in a small quantity of diet. The radioactivity reached a plateau value of 0.05 µg equivalents per g in the eggs 8 days after the start of dosing. Most of the residues were found in the yolk.

After about 8 daily doses of the parent compound the elimination of cypermethrin and/or its degradation products reached a balance in eggs with the daily uptake. About 30% of the radioactive residues in the yolk were unchanged cypermethrin; about 60% of the radioactive residues were closely associated with neutral ipids and phosphatidyl cholines cholines and could not be identified. 95% were eliminated with the faeces.

Concerning the residues in the respective tissues of this study it must be added that the residues were too low to allow for a characterisation. An accumulation of residues in muscle or the other tissues could not be observed. The highest residues were found in the liver of the animals (mean value of 4 animals: 0.37 µg/g). Only 0.05 µg/g of these was accounted for as cypermmethrin.

The values reported under [158)] in Table 82 are also based on a ^{14}C-study. The residues of deltamethrin detected in yolk were markedly higher than in egg albumen. The residues peaked 48 hours after the last treatment. Residues derived from the geminal diemthyl portion of the molecule tended to be higher than those from the benzyl moiety. The residues in the respective albumen dissipated much faster than in the yolk.

In this study low residues were found in muscles (traces to 0.21 mg/kg over the entire observation period of the study); see Table 82b.

The residue of cis- and trans-permethrin in the total egg reported under [161)] in Table 83 are the respective highest level occurring 3 to 4 days after the last dosing. The residue leNels are based on the unmetabolised parent compounds. The cis- and trans-

Table 83a. ^{14}C Permethrin Equivalents in Tissues and Organs 9 Days after Initiating a Treatment Schedule Consisting of Three Daily Doses of [acid-^{14}C]- or [alcohol-^{14}C] trans- or -cis-Permethrin at 10 mg/kg for Each Dose

Sample	[^{14}C] permethrin equivalents, ppm[a]			
	Trans		Cis	
	Acid	Alc	Acid	Alc
fat	0.21	0.18	1.03	1.36
skin	0.05	0.06	0.47	0.41
liver	0.14	0.08	0.27	0.20
kidney	0.31	0.24	0.34	0.25

[a] Average of two or three analyses on composite samples from two or three hens for each labelled preparation;
Table 83 is according to L. C. Caughan et al. [161)]

Table 83b. Residues of pyrethroids in chickens

Compound	Formulation	Days after last application	Country[1]	Residue levels (mg/kg) at harvest or at days after last application	Dosage a.i. (kg/ha)	Remarks (Ref.)
Permethrin	—	3 to 4 days after the last dose	USA	1.2 cis-permethrin and 0.3 trans-permethrin in whole egg	10 mg/kg on each of 3 consecutive days	161)
Cypermethrin, Deltamethrin Fenvalerate	—	4—5 days after oral administration	EG	0.21 in egg yolks[a] 0.27 in egg yolks 0.32 in egg yolks	10 mg/kg body weight, oral admin. of each of the 3 compounds in mixture	159)

[1] EG = Egypt;
[a] for further details see text

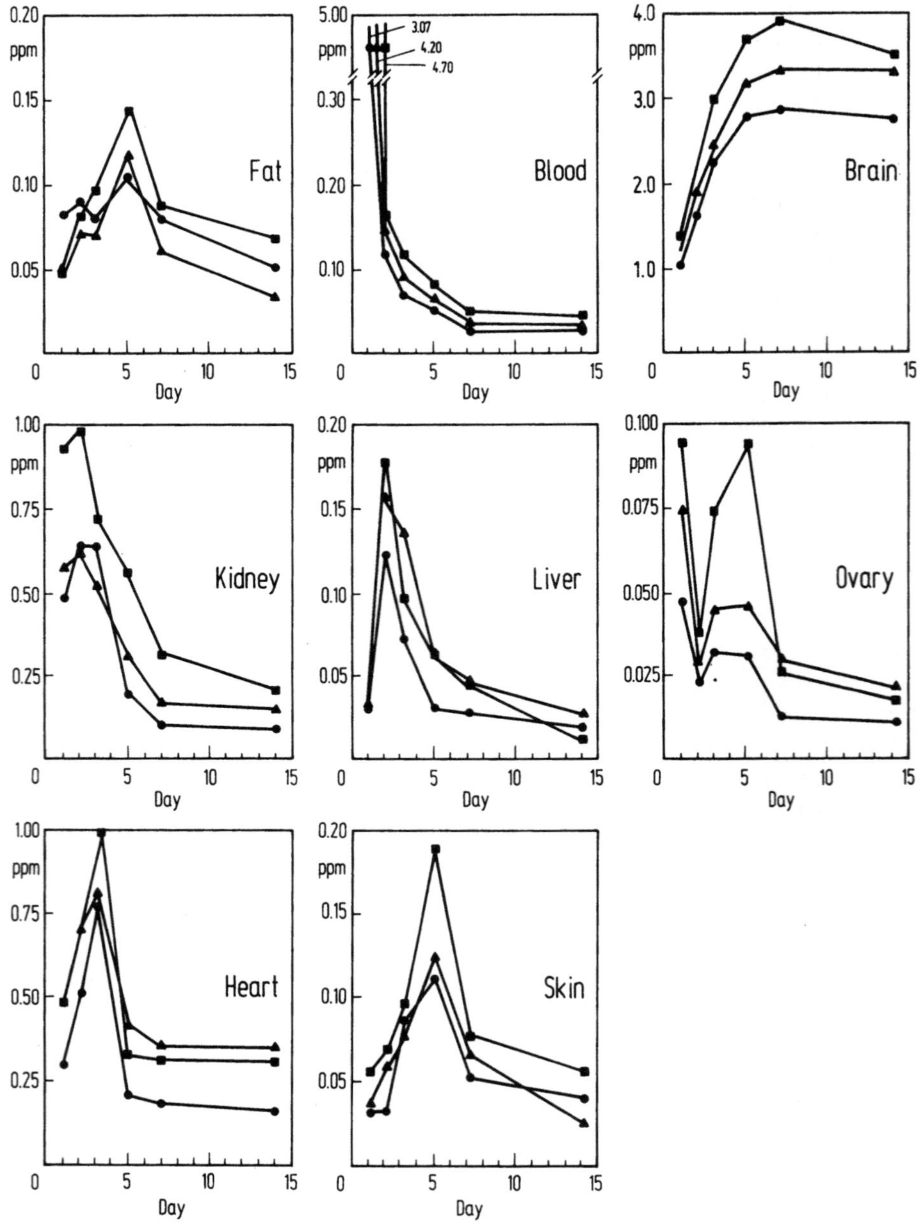

Fig. 9. Pyrethroid levels in tissues and organs of chickens treated orally at 10 mg/kg (mean for three chickens with individual results averaging 10—15% deviation from the mean): ■, fenvalerate; △, cymermethrin; ●, deltamethrin (according to M. A. Saleh et al. [159])

Table 84a. Radiocarbon residues in the white of eggs from laying hens treated with a single oral dose of [carbonyl-^{14}C]- or [furyl-^{14}C]- trans-, or of [carbonyl-^{14}C]- or [furyl-^{14}C]-cis-resemethrin

Days after treatment	Radiocarbon residue concentration[a] (expressed as μg resmethrin equivalents g^{-1})			
	Trans-Isomer		Cis-Isomer	
	[carbonyl-^{14}C]	[furyl-^{34}C]	[carbonyl-^{14}C]	[furyl-^{14}C]
1	0.16 (±0.06)	0.02 (±0.00)	0.34 (+0.05)	0.03[b]
2	0.02 (± <0.02)	0.02 (± <0.02)	0.08 (±0.03)	0.08 (± <0.02)
3	0.03 (±0.02)	0.02 (± <0.02)	0.04 (± <0.02)	(<0.02
4	<0.02	<0.02	0.04 (± <0.02)	<0.02
5	<0.02	<0.02	0.03 (± <0.02)	<0.02
6	<0.02	0.02 (±0.02)	0.02 (± <0.02)	<0.02
7	<0.02	0.02 (±0.02)	0.02 (± <0.02)	<0.02
8	<0.02	<0.02	0.02 (± <0.02)	<0.02
9	<0.02	<0.02[b]	<0.02	<0.02
10	<0.02	<0.02	<0.02	0.04 (±0.04)
11	<0.02	<0.02	<0.02	0.02 (± <0.02)
12	<0.02	0.02 (±0.02)	<0.02	<0.02
13	<0.02	<0.02	<0.02[b]	<0.02
14	<0.02	<0.02	<0.02	<0.02[b]

[a] Each value given is the average residue (±s.d.) found in the egg whites of two or three hens. A sensitivity limit of 0.02 μg g^{-1} was established for the combustion method used;
[b] Only one egg was laid at this collection interval

Table 84b. Radiocarbon residues in the yolk of eggs from laying hens treated with a single oral dose of [carbonyl-^{14}C]- or [furyl-^{14}C]- trans-, or of [carbonyl-^{14}C]- or [furyl-^{14}C]-cis-resmethrin

Days after treatment	Radiocarbon residue concentration[a] (expressed as μg resmethrin equivalents g^{-1})			
	Trans-Isomer		Cis-Isomer	
	[carbonyl-^{14}C]	[furyl-^{14}C]	[carbonyl-^{14}C]	[furyl-^{14}C]
1	0.07 (± <0.04)	0.04	0.05 (±0.04)	0.04[b]
2	0.17 (±0.04)	0.28 (±0.06)	0.18 (±0.05)	0.17 (±0.08)
3	0.25 (±0.04)	0.53 (±0.14)	0.28 (±0.11)	0.58 (±0.19)
4	0.28 (± <0.04)	0.62 (±0.15)	0.39 (±0.04)	0.58 (±0.16)
5	0.30 (±0.06)	0.60 (±0.15)	0.36 (±0.04)	0.57 (±0.13)
6	0.27 (± <0.04)	0.53 (±0.19)	0.30 (± <0.04)	0.51 (±0.14)
7	0.23 (± <0.04)	0.36 (±0.16)	0.25 (±0.05)	0.51 (±0.12)
8	0.04 (± <0.04)	0.15 (± <0.03)	0.13 (±0.06)	0.25 (±0.08)
9	<0.04	0.07[b]	0.06 (± <0.04)	0.10 (±0.05)
10	<0.04	0.04 (±0.04)	<0.04	0.11 (± <0.04)
11	<0.04	<0.04	<0.04	0.06 (± <0.04)
12	<0.04	<0.04	<0.04	0.05 (± <0.04)
13	<0.04	<0.04	<0.04[b]	0.04 (± <0.04)
14	<0.04	<0.04	<0.04	<0.04[b]

[a] Each value given is the average residue (±s.d.) found in the egg yolks of two or three hens. A sensitivity limit of 0.4 μg g^{-1} was established for the combustion method used;
[b] Only one egg was laid at this collection interval

Table 84c. (cont.) Distribution of ^{14}C-labelled residues in the tissues of laying hens, treated 12 h previously with a single oral dose of [carbonyl-^{14}C]- or [furyl-^{14}C]-trans-resmethrin, or of [carbonyl-^{14}C]- or [furyl^{14}C]-cis-resmethrin (10 mg/kg)

Tissue	Residue concentration in tissue[a] (expressed as µg resmethrin equivalents g^{-1} wet wt)			
	Trans-Isomer		Cis-Isomer	
	[carbonyl-^{14}C]	[furyl-^{14}C]	[carbonyl-^{14}C]	[furyl-^{14}C]
Kidney	3.1 (±1.3)	10.5 (±5.6)	7.2 (±5.2)	3.9 (±1.7)
Liver	2.7 (±2.0)	3.0 (±1.7)	3.9 (±1.8)	2.8 (±0.6)
Blood	1.3 (±1.1)	1.9 (±1.0)	4.0 (±3.1)	0.7 (±0.2)
Lungs	0.4 (±0.2)	1.5 (±1.0)	2.2 (±1.5)	0.5 (±0.3)
Skin	0.2 (± <0.1)	1.1 (±0.4)	1.5 (±1.0)	0.5 (±0.1)
Heart	0.2 (±0.2)	0.6 (±0.3)	1.5 (±1.1)	0.4 (± <0.1)
Gizzard	0.2 (±0.1)	0.8 (±0.8)	1.2 (±0.8)	0.2 (± <0.1)
Muscle (breast)	0.2 (±0.1)	0.2 (± <0.1)	0.7 (±0.5)	<0.1
Muscle (thigh)	0.3 (±0.2)	0.3 (±0.1)	0.8 (±0.6)	0.2 (± <0.1)
Fat	0.1 (± <0.1)	0.3 (±0.1)	0.4 (±0.2)	0.7 (±0.1)
Ovarian yolk[b]	0.2 (± <0.1)	0.3 (±0.1)	0.6 (±0.2)	0.7 (±0.3)

[a] The values (X ± s.d.) given are from three hens treated with the individual resmethrin preparations. The determinations were made by oxygen combustion analysis;
[b] Refers to the cluster of maturing follicles surrounding the ovary;
Table 84 is according to R. J. Christopher [162]. For further details see text

permethrin residues found in the fat depots of the laying hens in the course of this study were at levels of 0.93 mg/kg and 0.15 mg/kg resp. at 7 days after the last dosing.

According to [159] in Table 83 only 0.05 to 0.55 mg/kg of residues of the 3 pyrethroids cypermethrin, deltamethrin and fenvalerate were detected in the total egg with a peaking of each of the 3 pyrethroids in mixture occurring 4 to 5 days after oral administration of 10 mg per kg of body weight. Fenvalerate was the most persistent of the 3 active ingredients in eggs and resulted in the highest residues.

This applies also to the residue situation of the 3 pyrethroids in different tissues: fenvalerate was more persistent in all tissues than cypermethrin and deltamethrin (see also Fig. 9). Of interest is the finding that the residues of the parent compound in the brains were at much higher levels and much more persistent than those in the other investigated tissues.

Further residue data of pyrethroids in eggs and in tissues of laying hens can be taken from Tables 84 to 86. Thus, R. J. Christopher [162] reported about residues after oral administration of 10 mg/kg of cis- and trans resmethrin being ^{14}C-labeled in the alcohol or acid moiety. In this case more than 90% of the radioactivity were recovered in the excreta within 24 hours after administration. In eggs the highest levels 1 day after treatment were in the albumen and 4–5 days after treatment in the yolk. In tissues and in the excreta there were no great differences in the residue situation with regard to the isomers or the label position respectively. The highest levels were found in liver and kidney.

J. B. Moore [173)] reported about pyrethrin residues in eggs and in chicken tissues resulting from dipping chicken. The synergist piperonyl butoxide was also included in the investigations. Dipping of the birds is of course an unusual procedure resulting in a very high dose. However, this was intentionally applied in order to obtain measurable residues. The dipping was performed twice a week for 7 weeks. Even this unrealistically high number of treatments did not result in unusually high residues.

H. E. Braun and G. A. Surgeoner [175)] determined permethrin residues in tissues and egg yolk and egg white of hens (see Table 86). Significant levels were only found in fatty tissues after 42 days. While neither the cis-nor the trans-isomer were detectable in egg white during the entire observation period, permethrin residues occurred in egg yolk on the 3rd day after treatment at an average level of 0.01 mg/kg. The highest residue levels were measured 7 days after treatment (0.025 mg/kg). Until the 21st day after application these levels declined to traces or even undetectable levels (<0.001 mg/kg), respectively. This study also revealed that the trans-isomer is more rapidly eliminated and/or metabolised than the cis-isomer. When taking an assumed permissible level of 0.05 mg/kg of permethrin in edible chicken tissues as a basis, this level was exceeded only 1 day after treatment in the body fat, while it took 42 days for it to decline below the 0.05 m/kg level in the uropygium.

2.2.6 Household, Hygiene, Stored Product Protection and Effects of Processing

In the *household and hygiene field of application* pyrethroids are applied against respective pests such as cockroaches or flour moths.

Blaß and Riegner [166)] as well as Inkmann-Koch and Elbert [178)] presented results of studies concerning the contamination of selecred foodstuffs during and after the treatment of rooms with fenfluthrin [166)] and cyfluthrin [178)]. In their investigations, Blaß and Riegner used electric vaporizer mats which contained 50 mg fenfluthrin. At a temperature of the respective electric vaporizer of 90 °C, the active ingredient was evaporated in a kitchen for 8 or 24 hours, respectively.

Table 85a. Analysis of chicken tissues for pyrethrins, lower concentration

Tissue	Treatment[a]	Weeks	G. tissue/50 ml	Peak height (mm)		Pyrethrins (p.p.m.)
Body fat	TI	7	2.5	4	24.2	1.594
		11	2.5	4	2.4	0.158
Skin	TI	7	1.25	4	41.5	5.468
		11	1.25	4	4.5	0.712
Muscles	TI	7	25	4	20.8	0.137
		11	25	4	2.2	0.015
Liver	TI	7	25	4	12.3	0.081
		11	25	4	3.3	0.021
Gizzard	TI	7	25	4	20.4	0.134
		11	25	4	5.2	0.034

[a] See text

Table 85b. Analysis of chicken tissues for pyrethrins, higher concentration

Tissue	Treatment[a]	Weeks	G. tissue/ 50 ml	Injection (µl)	Peak height (mm)	Pyrethrins (p.p.m.)
Body fat	TII	7	2.5	4	73.3	4.829
		11	2.5	4	41.1	2.708
Skin	TII	7	0.125	4	45.1	59.420
		11	0.5	4	22.4	7.378
Muscles	TII	7	25	4	61.0	0.402
		11	25	4	24.6	0.162
Liver	TII	7	25	4	19.8	0.130
		11	25	4	3.9	0.024
Gizzard	TII	7	25	4	92.8	0.611
		11	23	4	29.3	0.193

[a] See text;
TI: emulsion containing 0.085% pyrethrins and 0.85% piperonyl butoxide;
TII: 10-fold amounts of TI (for further details see text)

Table 85c. Analysis of eggs for pyrethrin I resulting from dipping chickens in pyrenone, two concentrations

Treatment	Weeks	Injection (µl)	Peak height (mm)	Pyrethrins in eggs (p.p.m.)
TI	1	2	1.2	0.0097
TI	2	2	7.5	0.0608
TI	3	2	11.2	0.0908
TI	4	2	17.0	0.1379
TI	5	2	23.4	0.1898
TI	6	2	20.6	0.1671
TI	7	2	27.8	0.2255
0	8	2	10.8	0.0876
0	9	2	6.0	0.0487
0	10	2	5.4	0.0438
0	11	2	3.3	0.0268
TII	1	2	24.7	0.2003
TII	2	2	41.3	0.3350
TII	3	2	52.1	0.4225
TII	4	2	54.0	0.4380
TII	5	2	65.1	0.5280
TII	6	2	72.3	0.5864
TII	7	2	80.4	0.6521
0	8	2	45.2	0.3666
0	9	2	29.1	0.2360
0	10	2	25.0	0.2028
0	11	2	20.8	0.1687

Table 85d. Analysis of eggs for piperonyl butoxide resulting from dipping chickens in pyrenone, two concentrations

Treatment	Weeks	Injection (μl)	Peak height (mm)	Pyrethrins in eggs (p.p.m.)
TI	1	1	3.3	0.362
TI	2	1	7.1	0.779
TI	3	1	8.4	0.922
TI	4	1	12.3	1.349
TI	5	1	14.1	1.547
TI	6	1	16.4	1.799
TI	7	1	16.1	1.767
0	8	1	14.4	1.580
0	9	1	9.7	1.064
0	10	1	6.0	0.658
0	11	1	5.0	0.549
TII	1	1	23.5	2.579
TII	2	1	31.6	3.468
TII	3	1	34.5	3.787
TII	4	1	36.5	4.005
TII	5	1	38.5	4.225
TII	6	1	37.3	4.093
TII	7	1	39.8	4.367
0	8	1	32.4	3.555
0	9	1	29.6	3.248
0	10	1	25.3	2.776
0	11	1	20.5	2.249

Table 85 is according to J. B. Moore [173].
TI: emulsion containing 0.085% pyrethrins and 0.85% piperonyl butoxide;
TII: 10-fold amounts of TI (for further details see text)

The foodstuffs exposed in opened or unopened original packing were ultra-heat treated milk, hard sausage and rye bread.

Depending upon the kind of foodstuff used, samplings were made between 7 hours and 14 days after treatment.

The results are summarised in Table 87.

It can be taken from Table 87 that residues were detectable in milk only in the open 48-hour samples, the values being in the range of 30 μg/kg. Residues of fenfluthrin were not detectable in hard sausage (< 10 μg/kg), while in rye bread residues occurred between < 3 and 908 μg/kg. In this case the residues "peaked" in the open packings of the 24-hour samples. In the resepective 48-hour samples the values declined again by the factor of 30 to 50.

Inkmann-Koch and Elbert [178] determined residues of cyfluthrin in butter, milk, bread crumbs, flour, sugar and butter biscuits after a space spray of Baythroid H WP10 as recommended in practice. During the application (20 mg a.i./m^2 surface area to be treated) the foodstuffs were factory-packed and covered with polyethylene sheet. The actual exposure of the foodstuffs followed after a waiting period of 6 hours and ventilation of the room. The duration of the exposure was 0–4 days (for milk and butter and 0–16 days resp. for the other foodstuffs. The determination of the residues was per-

Table 86. Residues of permethrin (ppm) in tissues of hens treated with Ectiban® at 20 mg a.i./bird

Average[a] residue (ppm) at indicated time post-treatment

Post-Treatment (days)	Permethrin (isomer)	Liver	Internal Gizzard	Fat	Uropygium (Tail)	Breast Muscle	Egg Yolk	Egg White
1	cis	.009	—[b]	.043	.090	.010	—	—
	trans	.013	—	.072	.094	.012	—	—
	total	.022	—	.115	.184	.022	—	—
3	cis	.002	—	.012	.056	.005	.006	—
	trans	.003	.001	.013	.046	.007	.004	—
	total	.005	.001	.025	.102	.012	.010	—
7	cis	.001	.001	.019	.082	.006	.020	—
	trans	.002	.001	.015	.089	.008	.005	—
	total	.003	.002	.034	.171	.014	.025	—
14	cis	—	—	.009	.025	—	—	—
	trans	—	—	.006	.015	—	—	—
	total	—	—	.015	.040	—	—	—
21	cis	—	—	.013	.043	.002	.003	—
	trans	—	—	.005	.049	.003	.002	—
	total	—	—	.018	.092	.005	.005	—
42	cis			.011	.019			
	trans			.006	.027			
	total	NA[c]	NA[c]	.017	.046	NA	NA	NA

[a] Average of 3 replicates;
[b] Not detectable (<0.001 ppm);
[c] NA — not analysed;
Table 86 is according to H. E. Braun and G. A. Surgeoner [175])
For further details see text

formed on substrates having been exposed loosely, factory-packed or in their opened packing.

In all cases residues of cyfluthrin were not detectable (limits of detection between 30 and 50 μg/kg).

Result: residues of cyfluthrin in foodstuffs are not to be expected after application of cyfluthrin in the house-hold and hygiene field if applied as recommended in practice.

Numerous publications dealing with the use of pyrethroids for *stored product protection* and the residue situation in the respective foodstuffs resulting from this use have already been quoted [84,118,119,120,124,125,126,127,129,131,133,134,135]). As pyrethroids are mainly used for stored product protection in cereals, the test results are shown and discussed in Chapter 3 "Cereals and grain products".

The effect of processing must be investigated in order to determine the actual pyrethroid residue intake by the consumer from the foodstuffs after having been prepared and being ready for consumption. As a rule it is to be expected that washing, cooking, frying as well as other processing techniques used in the kitchen will result in a reduction of the residues. In fact, an aggravation of the residue situation is only conceivable

Table 87. Residues in foodstuffs after the use of fenfluthrin electric vaporizer mats; according to W. Blaß and K. Riegner [166)]

	Kitchen with							
	8-hour evaporation				24-hour evaporation			
Matrix	T	open	T	unopened	T	open	T	un-opened
Ultra-heat treated milk	7 h	n.d.	2 d	n.d.	7 h	n.d.	2 d	n.d.
	24 h	n.d.	7 d	n.d.	24 h	n.d.	7 d	n.d.
	48 h	31	14 d	n.d.	48 h	27	14 d	n.d.
hard sausage	7 h	n.d.	2 d	n.b.	7 h	n.d.	2 d	n.b.
	24 h	n.d.	7 d	n.b.	24 h	n.d.	7 d	n.b.
	48 h	n.d.	14 d	n.d.	48 h	n.d.	14 d	n.d.
rye bread	4 h	31	4 h	appr. 5	4 h	n.d.	4 h	n.d.
	24 h	908	24 h	n.d.	24 h	826	24 h	12
	48 h	19	48 h	appr. 8	48 h	26	48 h	18

All values are given in µg/kg;
T: time of sampling;
open: opened packing;
unopened: unopened original packing;
n.d.: not detectable, as below the limit of detection of:
a) milk = 5 ppb,
b) bread = 3 ppb,
c) sausage = 10 ppb;
n.b.: not determined

if residues of thermostable active ingredients are present which cannot be washed off and if at the same time greater losses of water result from the processing in the kitchen.

Several of such processing studies have already been quoted and evaluated [53,73,78,118,119,120,129,135)].

From Table 58 ("Residues of cyfluthrin in wheat") respective residue data in wheat grains treated post-harvest and in products processed therefrom, can be taken [81)]. In pollard and bran the residues were higher than in grain while in finely sifted wheat flour and white bread from this they were lower than the reported levels in the grain.

In a pure processing study Celino and Magallona [176)] investigated the behaviour of permethrin residues on Pole Sitao Beans (Phaseolus vulgaris L.) after washing and cooking of the beans. Washing reduced the permethrin content by 20%, washing and cooking by approximately 60% of the content in unwashed material which contained 27.7 µg/25 g of sample.

T. M. Chen [177)] investigated the residue situation of permethrin on tea leaves including the residues in a respective tea infusion. By the infusion procedure only 3.9% to 4.7% of the permethrin residues were transferred from the levves into the made tea.

Table 88. Examples of Maximum Residue Limits for Residues of Pyrethroids (Source: P. R. Bennett, Health and Welfare, Ottawa, Canada; "National Pesticide Residue Limits in Foods" (1988))

Country	Substance	Remarks	Crop or Food	Max. Quantity mg/kg
Fed. Rep. of Germany	Fenvalerate		Meat and meat products,	0.05
			milk and milk products	0.01[a]
	Fenvalerate		Berries, pome fruits, stone fruits (except for plums)	2.0
			Grapes	1.0
			Plums	0.5
			Potatoes, maize, rape, sug. beets	0.05
			other vegetable foodstuffs	0.02
	Piperonyl butoxide	Synergist	Cereals	10.0
			Vegetables except root vegetab., fruits, spices, raw coffee, tea tea-like products, oilseed	3.0
			Other vegetable foodstuffs	0.5
	Pyrethrins		Cereals	3.0
			Vegetables, fruits	1.0
			Other vegetable foodstuffs	0.5
	Tetramethrin		Cereals, legumes	0.1
India	Pyrethrins		Foodgrains	1.5
			Milled foodgrains	0.5
			Fruits and vegetables	1.0
Japan	Cypermethrin		Tea	30.0
			Exocarp of summer orange, vegetables	5.0
			Fruits (except exocarp of summer orange)	2.0
			Sugar beet	0.2
			Potato, etc.	0.05
	Cycloprothrin		Rice	0.1

Sum of pyrethrins I and II and other structurally related insecticidal ingredients of pyrethrum

	Fenvalerate	Exocarp of summer orange	20.0
		Fruits (except exocarp of summer orange)	1.0
		Vegetables, sugar beet	0.5
		Pulses	0.1
		Potato etc.,	0.05
	Flucythrinate	Tea	15.0
		Exocarp of summer orange	5.0
		Fruits (except exocarp of summer orange)	2.0
		Vegetables, sugar beet	0.5
Japan (cont.)	Fluvalinate	Tea	10.0
		Exocarp of summer orange	5.0
		Fruit (except exocarp of summer orange), vegetables	0.5
	Permethrin	Tea	20.0
		Exocarp of summer orange	15.0
		Fruits (except exocarp of summer orange)	5.0
		Vegetables	3.0
		Potato, etc.	0.2
	Tralomethrin	Tea	10.0
		Exocarp of summer orange	2.0
		Fruits (except exocarp of summer orange), vegetables	0.5
		Potato, etc.	0.05
USA	Allethrin	Post harvest use of apples, blackberries, blueberries (huckleberries), boysenberries, cherries, crabapples, currants, dewberries, figs, gooseberries, grapes, guavas, loganberries, mangoes, muskmelons, oranges, peaches, pears, pineapples, plums (fresh prunes), raspberries, tomatoes	4.0
		Post harvest use on grains of barley, corn, sorghum, milo, oats, rye, wheat Jerusalem artichokes, beets,	2.0
		Chinese cabbage, carrots, celery, chickory, corn, endive, escarole, garlic, leeks, onions, parsley, potatoes, salsify, shallots, grain sorghum, sugar beets, milo sorghum, spinach, sweet potatoes	Exempt

Table 88. (Continued)

Country	Substance	Remarks	Crop or Food	Max. Quantity mg/kg
	Cyfluthrin		Tomato pomace (dry)	5.0 (tempor. tol.)
			Tomato pomace (wet)	2.0 (tempor. tol.)
			Cottonseed hulls and oil	2.0
			Cottonseed	1.0
			Meat, fat and meat by-products of cattle, goats, hogs, horses and sheep	0.05
			Milk	0.01
USA (cont.)	Cypermethrin		Pecans	0.05[b]
			Lettuce	10.0*)[b]
	Deltamethrin		Cottonseed oil	5.0
	Fenvalerate		Corn fodder and forage	50.0
			Dried apple pomace (animal feed), sugarcane bagasse, turnip tops	20.0
			Almond hulls	15.0
			Peaches, stone fruits	10.0
			Radish tops	8.0
			Milk fat	7.0
			Blueberries, caneberries, currants, elderberries, gooseberries, huckleberries	3.0
			Pears, broccoli, sunflower hulls, sugarcane	2.0
			Peanut hulls, soybean hulls, peas, eggplants, sunflower seeds	1.0
			Fat, meat and meat by-products of cattle, goats, hogs, horses and sheep	1.5
			Cauliflower, carrots, turnip roots	0.5
			Milk, radish roots	0.3
			Dried beans and peas	0.25
			Dried apple pomace, artichokes, filberts, English walnuts, almonds	0.2
			Corn sweet kernel with cob	0.1
			Okra	0.1[c]
			Corn grain (field)	0.02

USA (cont.)	Flucythrinate		Apples, fat of cattle, goats, hogs, horses and sheep, sweet corn cannery waste	1.0
			Apple pomace (dry)	10.0
			Cottonseed, meat and meat by-products of cattle, goats, hogs, horses and sheep, milk	0.1
			Milk fat, lettuce (head), cabbage	2.0
			Pears, corn grain	0.05
			Cottonseed oil	0.02
			Corn fodder and forage	3.0
	Fluvalinate		Cottonseed	0.1
			Meat, fat and meat by-products of cattle, goat, hogs, horses, poultry, and sheep; milk; eggs	0.01
			Cottonseed hulls	0.3
			Cottonseed oil (crude and refined)	1.0
	MGK 264	Synergist	Milk fat and fat of cattle, goats, hogs, horses and sheep	0.3
			Processed foods	10.0
USA (cont.)	Permethrin		Corn fodder and forage	60.0[d]
			Spinach, almond hulls, lettuce head, turnip greens, collards	20.0[d]
			Mushrooms, cabbage	6.0[d]
			Peaches, celery, watercress	5.0[d]
			Artichokes	10.0[d]
			Milk fat (reflecting 0.25 ppm in whole milk)	3.75[d]
			Meat by-products of hogs, pears, cantaloupes, cherries	3.0[d]
			Kiwi fruit, fat of cattle, goats, hogs, horses, sheep, tomatoes, pumpkins	2.0
			Meat by-products of cattle, goats, horses, sheep, broccoli, avocados, Brussels sprouts, cauliflower, horseradish, asparagus, turnip roots	1.0[d]
			Papayas	1.0[e]
			Meat of cattle, goats, hogs, horses, sheep	0.15[d]
			Corn sweet (kernel plus cob with husks removed), pistachios	0.1[d]

Table 88. (Continued)

Country	Substance	Remarks	Crop or Food	Max. Quantity mg/kg
			Cottonseed	0.5[d]
			Eggs, milk and meat, fat and meat by-products of cattle, goats, hogs poultry, sheep, almonds, apples, soybeans, potatoes, corn grain (field and pop), filberts, walnuts	0.05[d]
USA (cont.)	Piperonyl butoxide	Synergist	Barley, bird seed mixtures, buckwheat, corn, rice, rye, wheat	20.0[f]
			Milled fractions from cereal grains, dried foods (bag cloth & paper cloth), dried feed (cotton bags & paper bags), processed food	10.0[f]
			Almonds, apples, beans, blackberries, blueberries, (huckleberries), boysenberries, cherries, cocoa beans, copra, cottonseed, crabapples, currants, dewberries, figs, flaxseed, gooseberries, grain sorghum, grapes, guavas, loganberries, mangoes, muskmelons, oats, oranges, peaches, peanuts (determined on the nuts with shell removed), pears, peas, pineapples, plums, (fresh prunes), raspberries, tomatoes, walnuts	8.0[f]
			Meat, fat and meat by-products of poultry	3.0
			Eggs	1.0
			Potatoes, sweet potatoes	0.25[f]
			Milk fat	0.25
			Meat, fat and meat by-products of cattle, goats, hogs, horses and sheep	0.1 (negligible residues)
USA (cont.)	Pyrethrins		Barley, birdseed mixtures, buckwheat corn rice, rye, wheat	3.0[g]
			Almonds, apples, beans, blackberries, blueberries (huckleberries), boysenberries, cherries, cocoa beans, copra, crabapples, currants, dewberries, dried foods (from cloth bags and from paper bags), dried feeds (from cotton bags and from paper bags), processed food, figs, flaxseed, gooseberries, grain sorghum, grapes,	1.0[g]

		grapes, guavas, loganberries, mangoes, milled fractions from cereal grains, muskmelons, oats, oranges, peaches, peanuts, (determined on the nuts with shell removed), pears, peas, pineapples, plums (fresh prunes), rasp-berries, tomatoes, walnuts	
		Milk fat	0.5 (negligible residues)
		Meat, fat and meat by-products of poultry	0.2
		Eggs	0.1 (negligible residues)
		Potatoes, sweet potatoes	0.05[g]
		Cottonseed	0.05[h]
		Meat, fat and meat by-products of cattle goats, hogs, horses, speep; milk	0.01[h]
USA (cont.)	Resmethrin	Food items	3.0
	Tralomethrin	Soybeans	0.05[i]
		Cottonseed oil cottonseed 0.02	0.2
U.S.S.R.	no tolerance levels for pyrethroid residues given		

[a] In case of milk products, except for milk fat, based on the weight of the milk used for products;
[b] incl. its metabolites "DCVA" and "3-PB acid";
[c] limited to Florida;
[d] incl. its metabolites, 3(2,2-dichloroethenyl)-2,2-dimethylcyclopropane carboxylic acid (DCVA) and (3-phenoxybenzyl)-methoanol (3-PBA);
[e] limited to Florida;
[f] postharvest application;
[g] postharvest application;
[h] expires August 30, 1991;
[i] incl. its metabolite (S)-α-cyano-3-phenoxybenzyl (1R, 3R)-cis, trans-2,2-dimethyl-3(2,2-dibromovinyl) cyclopropanecarboxylate;

Comments: Table 88 shall only give an overview of the MRLs in different countries which were selected arbitrarily. For further MRLs for pyrethroids established in other countries we refer to the above-mentioned publication of P. R. Bennett

3 Final Remarks

In the meantime the broad-spectrum synthetic pyrethroids have partly replaced the insecticidal organic phosphates and carbamates. Pyrethroids act on the sodium channels of nerve membranes of insects and have thus a similar mode of action as the insecticidal chlorinated hydrocarbons.

Pyrethroids are considered to be of low toxicity, relatively photostable and yet biodegradable. The normal degradational pathway involves an oxidation and ester cleavage with subsequent conjugation under formation especially of glycosides. Animals eliminate pyrethroids mainly with the faeces and the urine.

Despite a relatively high stability of the compounds, no persistent residues are formed in animals, i.e. there is practically no bioaccumulation of these substances.

The relatively high fish toxicity is a disadvantage of these compounds. Depending upon the examined kind of fish the toxicity is in the order of about 0.2 to 30 µg/l.

With the exception of fluvalinate, pyrethroids are considered to be fairly highly toxic to honey bees. However, some of these substances seem to have a repellent effect upon bees. On account of the low rates applied in insecticidal treatments with pyrethroids, the relatively high bee toxicity of these substances is, however, at least partly compensated and they must therefore, no doubt, not be rated as less favourable than the insecticidal organic phosphates or carbamates which are applied at considerably higher rates.

In tropical and subtropical regions with their intense solar radiation and the heavy rainfalls, the relatively high photostability and the very low water solubility of these compounds is an advantage compared to the insecticidal organic phosphates and carbamates which should not be underestimated. The latter products must be applied much more frequently in those regions in order to assure that the pests are controlled successfully.

The residue data reported and partly evaluated in the present paper were mainly taken from scientific literature or elaborated in the laboratory of the author. A claim of completeness can, of course, not be made because it is to be assumed that numerous data are available to the responsible registration authorities or have been elaborated by the manufacturers without having been published so far.

Based on the respective residue tolerances (see also Table 88) and on the residue data being available so far, the residue situation of pyrethroids in or on food stuffs can be considered to be favourable.

Acknowledgement. My special thanks are expressed to Dr. W. Rabe for the translation of the text from German into English, as well as to Dr. H. Frehse and Dr. E. Möllhoff for a careful review of the manuscript and valuable suggestions.

4 References

1. Casida JE (1980) Environmental Health Perspectives 34: 189
2. Naumann K (1989) Structures, properties and chemistry of synthetic pyrethroid insecticides (in press) Springer Berlin Heidelberg New York; editors: G. Haug and H. Hoffmann
3. Elliot MA, Jones NF (1978) Chem. Soc. Rev. (London) 7: 473
4. Papadopoulou-Mourbidou E (1983) Residue Reviews 89: 179

5. Naumann K (1981) Chemie der synthetischen Pyrethroid-Insektizide (publ. in: R. Wegler: Chemie der Pflanzenschutz- u. Schädlingsbekämpfungsmittel, vol 7; Springer, Berlin Heidelberg New York
6. Watts RR, Storherr RW (1968) J. Assoc. Off. Anal. Chem. 51: 662
7. Watts RR, Storherr RW (1967) J. Assoc. Off. Anal. Chem. 50: 581
8. Storherr RW, Watts RR (1965) J. Ass. Off. Agric. Chem. 48: 1154
9. Blaß W (1970) Dissertation, Universität Münster
10. Specht W (1987) Methodensammlung zur Rückstandsanalytik der „DFG", 9th Supplement, Verlag Chemie, Weinheim, p 19
11. Blaß W (July 1987) LABO 18: 54, Verlag Hoppenstedt, Darmstadt
12. Thier HP, Frehse H (1986) Rückstandsanalytik von Pflanzenschutzmitteln; In: Analytische Chemie für die Praxis, G. Thieme, Stuttgart
13. Hemingway RJ (1984) Pure and Appl. Chem. 56: 1131
14. Dicke W, Thier HP (1986) Lebensmittelchem. Gerichtl. Chem. 40: 11
15. Desmarchelier JM (1976) J. Stored Prod. Res. 12: 245
16. Wing KD et al. (1978) J. Agric. Food Chem. 26: 1328
17. Wing KD, Hammock BD (1979) Experientia 35: 1619
18. George DA et al. (1976/77) ACS-Sympos. Series, vol 42, p 201 ISBN 0-8412-0368-7)
19. Anonymous (April 1986) Talstar, ein neues Pyrethroid mit akarizider Wirkung, BAYER AG, Pflanzenschutz Anwendungstechnik Beratung (unpublished)
20. Wagner K: BAYER-Report No. RA-355, April 21, 1981 (unpublished)
21. Wagner K: BAYER-Report No. RA-454, April 20, 1982 (unpublished)
22. Gronberg RR, Pfankuche LK: MOBAY Report No. 85886, June 15, 1983 (unpublished)
23. Shaw HR II et al.: MOBAY Report No. 86217, Nov. 14, 1983 (unpublished)
24. Shaw HR II et al.: MOBAY Report No. 85883, July 25, 1983 (unpublished)
25. Harbin AM et al.: MOBAY Report No. 85823, June 14, 1983 (unpublished)
26. Ishii Y, Ueyama I: Report No. 1179 (ESR), July 15, 1983 (NITOKUNO) (unpublished)
27. Riegner K: BAYER Report No. RA-377/Doc.-No.: 0545B, June 11, 1986 (unpublished)
28. Inkmann-Koch A: BAYER Report No. RA 40/Doc.-No. 1895I (1986) (unpublished)
29. Gronberg RR et al.: MOBAY Report No. 91954 (April 2, 1986) (unpublished)
30. Blaß W: BAYER Report No. RA-434, June 9, 1987, Doc.-No. 0708B (unpublished)
31. Khan SU et al. (1986) Chemosphere 15: 923
32. Krohn J et al.: BAYER, Werk Elberfeld, Laboratory Report of April 23, 1986: Optimierung der Enantiomerentrennung des Baythroids durch HPLC (unpublished report)
33. Chapman RA (1983) J. Chromatogr. 258: 175
34. Jacques A et al. (1984) Anal. Methods Pestic. Plant Growth Regul. 13: 9
35. Chapman RA, Harris CR (1979) J. Chromatogr. 174: 369
36. Braun HE, Stanek J (1982) J. Assoc. Offic. Anal. Chem. 65: 685
37. Sapiets A et al. (1984) Anal. Methods Pestic. Plant Growth Regul. vol 13, p 33
38. Forbes S, Dutton AJ (1985) Pestic. Sci. 16: 404
39. Sarna LP, Webster GRB (1985) Intern. J. Environ. Anal. Chem. 21: 179
40. Kadenczki L et al. (1985) Novenyvedelem. (Budapest) 21: 394
41. Akhtar MH (1982) J. Chromatogr. 246: 81
42. Agneessens R et al. (1984) J. High Res. Chrom. and Chrom. Comm. 7: 533
43. Vaysse M et al. (1984) Anal. Methods Pestic. Plant Growth Regul. 13; chap 3, DECIS
44. Meinard C et al. (1985) J. Chromatogr. 349: 109
45. Meinard C et al. (1985) J. Chromatogr. 349: 105
46. Takimoto J et al. (1984) Anal. Methods Pestic. Plant Growth Regul. 13: 69
47. Saleh MA et al. (1980) J. Agric. Food Chem. 28: 592
48. Baker PG, Bottomley P (1982) Analyst 107: 206
49. Weinmann WD, Nolting H-G, Siebers J (1985) Methodensammlung zur Rückstandsanalytik (DFG, 8th) Verlag Chemie, Weinheim, p 23
50. Bolygo E, Zabar F (1983) J. Ass. Off. Anal. Chem. 66: 1013
51. Blaß W, Riegner K: BAYER-Report No. 1697, May 12, 1982 (unpublished)
52. Hill BD et al. (1982) J. Agric. Food Chem. 30: 653
53. Spittler TD et al. (1982) J. Ass. Off. Anal. Chem. 65: 1106

54. Spittler TD et al. (1984) J. Ass. Off. Anal. Chem. 67: 824
55. Shell Co. (1984) Fenvalerate: Anal. Methods Pestic. Plant Growth Regul., vol 13, p 121
56. Papadopoulou-Mourbidou E (1985) Chromatographia 20: 376
57. Cordon C (1986) J. Agric. Food Chem. 34: 953
58. Neuhauser H (1982) Vet. Med. Rev., Heft 2, p 158
59. Leidy RB et al. (1986) Beiträge zur Tabakforsch. Intern. 13: 191
60. Fitch WL et al. (1984) Anal. Methods Pestic. Plant Growth Regul. 13: 79
61. MacPhee A et al. (1982) J. Environm. Sci. Health, B17: 183
62. Papadopoulou-Mourbidou E et al. (1983) J. Agric. Food Chem. 31: 629
63. Stan HJ, Mrowetz D (1983) J. Chromatogr. 279: 173
64. Blaß W et al. (1979) J. Chromatogr. 172: 67
65. Fehringer NV, Walters SM (1984) J. Assoc. Off. Anal. Chem. 67: 91
66. Sundaram KMS et al. (1985) J. Liquid Chromatogr. 8: 2607
67. Takimoto Y et al. (1984) Anal. Methods Pestic. Plant Growth Regul. 13: 133
68. Siltanen H et al. (1978) Pyrethrum Post 14: 65
69. Weinmann WD, Nolting HG, Siebers J (1985) Rückstandsanalytik von Pflanzenschutzmitteln, DFG, Supplement 8th, Verlag Chemie, Weinheim, p 22
70. Ando T et al. (1986) Agric. Biol. Chem. 50: 491
71. Desmarchelier JM (1980) J. Pestic. Sci. 5: 521
72. Amos R, Leahey JP (1986) Brit. Crop Protect. Conf., Proceedings vol 2, p 821
73. Rai S et al. (1980) Indian J. Ent. 42: 657
74. Siebers J, Nolting HG (1982) Nachrichtenbl. Deut. Pflanzenschutzd. 34: 166
75. Awasthi MD (1985) J. Food Sci. Technol. 22: 4
76. L'Hotellier M (1982) Deltamethrin Monograph, chapter 8, p 285 (Roussel Uclaf); ISBN 2-904125-01-9
77. Awasthi MD, Anand L (1984) Pestic. Environ. Proc. Natl. Semin., India, p 45
78. Jain HK et al. (1979) J. Ent. Res. 3: 212
79. Rai S et al. (1986) Indian J. Agr. Sci. 56: 667
80. Agnihotri NP et al. (1980) Indian J. Ent. 42: 717
81. Anonymous, Pestic. residues in food — 1986 FAO Plant Prod. Protect. Paper 78: "Evaluations 1986", Part I — Residues
82. George DA (1985) J. Ass. Off. Anal. Chem. 68: 1160
83. Greenberg RS (1981) J. Agric. Food Chem. 29: 856
84. Anonymous: WHO Pestic. Residues Series, No. 5; 1975 Evaluations of some pesticide residues in food; WHO, Geneva (1976)
85. Balinov I, Balinova A (1982) Gradinarska i Lozarska Nauka, Sofia, 19: 70
86. Grounds PR (1982) New Zealand J. Experim. Agric. 10: 425
87. Harris CR et al. (1977/78) Proc. Entomol. Soc. Ontario 108: 63
88. Cabras P et al. (1985) J. Agric. Food Chem. 33: 935
89. Awasthi MD (1986) Phytoparasitica 14: 231
90. Siltanen H, Mäkinen S (1982) Publ. of the State Inst. of Agric. Chem., No. 23, Helsinki
91. Anonymous: Pesticidrester i danske levnedsmidler 1982/83, Pestic. Residues in Danish Food; SL-Public. No. 107, May 1985
92. Agnihotri NP et al. (1983) J. Ent. Res. 7: 10
93. Awasthi MD, Anand L (1983) J. Ent. Res. 7: 55
94. Belanger A, Hamilton HA (1979) J. Environ. Sci. Health, B14: 213
95. Leahey JP, Carpenter PK (1980) Pestic. Sci. 11: 279
96. Dejonckheere et al. (1982) Pestic. Sci. 13: 351
97. Braun HE et al. (1982) Pestic. Sci. 13: 119
98. Lee PW (1985) J. Agric. Food Chem. 33: 993
99. Frank R et al. (1982) J. Econ. Entomol. 75: 560
100. Awasthi MD (1986) Indian J. Hortic. 43: 161
101. Khan SU et al. (1984) J. Agric. Food Chem. 32: 1141
102. Awasthi MD, Anand L (1985) Indian J. Agric. Chem. 18: 57
103. Gojbhiye VT et al. (1982) J. Ent. Res. 6: 137
104. Srivastava KP et al. (1987) Indian J. Agric. Sci. 57: 419
105. Srivastava KP et al. (1984) J. Ent. Res. 8: 1

106. Sukul P, Handa SK (1984) Indian J. Agric. Sci. 54: 315
107. Sukul P, Handa SK (1986) Indian J. Agric. Sci. 56: 605
108. East R, Holland PT (1984) Proc. 37th N.Z. Weed and Pest Control Conf., p 228
109. Hatfull RS (1983) J. Assoc. Publ. Analysts 21: 19
110. Belanger A et al. (1985) J. Econ. Entomol. 78: 463
111. Nicolson RS (1984) J. Assoc. Publ. Analysts 22: 51
112. Haffner K et al. (1984) Lebensmittelchem. Gerichtl. Chem. 38: 95
113. Haffner K, Matter L (1986) Gartenbauwissenschaft 51: 75
114. Haffner K et al. (1985) Gartenbauwissenschaft 50: 177
115. Chiba M (1978) J. Environ. Sci. Health B13: 261
116. Awasthi MD (1986) Indian J. Agric. Sci. 56: 378
117. Awasthi MD, Anand L (1984) Phytoparasitica 12: 135
118. Desmarchelier JM et al. (1980) J. Pesticide Sci. 5: 539
119. Dicke W et al. (1986) Lebensmittelchem. Gerichtl. Chem. 40: 63
120. Dicke W et al. (1987) Getreide, Mehl und Brot, Lebensmittelchem. Gerichtl. Chem. 41: 240
121. McEwen FL et al. (1986) Pestic. Sci. 17: 150
122. Frank R et al. (1987) J. Assoc. Off. Anal. Chem. 70: 1081
123. Westcott ND, Reichle RA (1987) J. Environ. Sci. Health, B22: 91
124. Noble RM, Hamilton DJ (1985) Pestic. Sci. 16: 179
125. Desmarchelier J et al. (1981) Pestic. Sci. 12: 365
126. Bengston M et al. (1983) Pestic. Sci. 14: 373
127. Walkden HH, Nelson HD (1959) U.S. Departm. Agric. Mark. Research Report 322, 48 pp
128. Noble RM et al. (1982) Pestic. Sci. 13: 246
129. Joia BS et al. (1985) J. Agric. Food Chem. 33: 618
130. Kocourek V et al. (1987) Z. Lebensm. Unters. Forsch. 185: 14
131. Webley DJ, Kilminster KM (1981) Pestic. Sci. 12: 74
132. Yamamoto I, Katsuda Y (1980) Pestic. Sci. 11: 134
133. Bengston M et al. (1983) Pestic. Sci. 14: 385
134. Bengston M et al. (1984) Pestic. Sci. 15: 500
135. Bengston M et al. (1987) Pestic. Sci. 21: 23
136. Gupta GP et al. (1984) J. Ent. Res. 8: 73
137. Agnihotri NP et al. (1986) Pesticides (Bombay) 20: 17
138. Agnihotri NP et al. (1986) Intern. J. Tropic. Agric. 4: 168
139. Gupta HCL, Kavadia VS (1984) Pesticides 18: 39
140. Sharma GC et al. (1986) Pesticides 20: 40
141. Sharma GC et al. (1984) Pesticides 18: 10
142. Agnihotri NP et al. (1987) Pesticides 21: 35
143. Khan P et al. (1987) Pakistan J. Sci. Ind. Res. 30: 132
144. Tappan WB et al. (1982) J. Econ. Entom. 75: 1143
145. Petersen JH, Jensen KG (1986) Z. Lebensm. Unters. Forsch. 182: 489
146. Gopal M et al. (1987) Pesticides 21: 33
147. Roberts TR, Standen ME (1981) Pestic. Sci. 12: 285
148. Ebing W (1985) Ber. üb. Landwirtsch. 198: 35
149. Lhoste J, L'Hotellier M (1982) Deltamethrin Monograph, chap. 9, p 321 (Roussel Uclaf); ISBN 2-904125-01-9
150. Harris CR et al. (1981) Can. Entomologist 113: 685
151. Talehar NS et al. (1983) J. Econ. Entomol. 76: 711
152. Hill BD (1983) J. Environ. Sci. Health B18: 691
153. Smith S, Willis GH (1987) Soil Science 144: 67
154. Schaefer CH et al. (1978) Proc. Pap. Annu. Conf. Calif. Mosq. Vector 46: 112
155. Agnihotri NP, Jain HK (1987) Pesticides 21: 36
156. Bahadir M et al. (1987) Chemosphere 16: 1311
157. Agnihotri NP et al. (1986) J. Ent. Res. 10: 147
158. Akhtar MH et al. (1985) J. Agric. Food Chem. 33: 610
159. Saleh MA et al. (1986) J. Agric. Food Chem. 34: 895
160. Hutson HD, Stoydin G (1987) Pestic. Sci. 18: 157
161. Gaughan LC et al. (1978) J. Agric. Food Chem. 26: 1374

162. Christopher RJ et al. (1985) Pestic. Sci. 16: 378
163. Braun E et al. (1985) Bull. Environ. Contam. Toxicol. 35: 61
164. Croucher A et al. (1985) Pestic. Sci. 16: 287
165. Akhtar MH et al. (1986) J. Agric. Food Chem. 34: 753
166. Blaß W, Riegner K: BAYER Report No. 1617 Jan. 15, 1982 (unpublished)
167. Frank R et al. (1984) Pestic. Sci. 15: 600
168. Wszolek PC et al. (1980) Bull Environ. Contam. Toxicol. 24: 296
169. Lubs M et al. (1983) Milchwissenschaft 38: 597
170. Hamann J et al. (1984) Milchwissenschaft 39: 408
171. Ridlen RL et al. (1984) J. Agric. Food Chem. 32: 1211
172. Byford RL et al. (1986) Bull. Environ. Contam. Toxicol. 37: 692
173. Moore JB (1970) Residue Reviews 33: 87
174. Wszolek PC et al. (1981) Bull. Environ. Contam. Toxicol. 27: 869
175. Braun HE, Surgeoner GA (1981) Proc. Annu. Workshop, Pestic. Residue Analysis 16: 7 (West Canada)
176. Celino LP, Magallona ED (1985) Philipp. Agric. 68: 525
177. Chen TM (1981) Acta Entomol. Sin. 24: 9
178. Inkmann-Koch A, Elbert D: BAYER Rep. No. 1715b, Aug. 1982 (unpublished)

Immunoassay for the Quantitation of Low Molecular Weight Compounds in Plants

E. W. Weiler
Lehrstuhl für Pflanzenphysiologie, Ruhr-Universität, Postfach 102148, 4630 Bochum, FRG

Immunoassay has become increasingly important for the quantitation of low molecular weight compounds in plant tissues. Requiring little sophisticated instrumentation, yet providing extreme versatility, sensitivity, specificity and simplicity of operation, monoclonal antibody-based enzyme immunoassays are among the most attractive analytical tools available today. In this review, a general introduction into the principles of immunoassay is given, followed by outlines for the practical realization of such assays. This introduction is meant to provide non-immunologists with the necessary background information rather than give an in-depth theory of immunoassay. Then, a compilation of available assays is given covering secondary plant metabolites, endogenous as well as synthetic plant growth regulators, insecticides, fungicides and herbicides. A survey on food and environmental analysis with special emphasis on mycotoxins concludes this chapter.

1 Introduction

Landsteiner's pioneering investigations on the antigenicity of small molecules [186], culminating in the development of the hapten concept, has provided the framework for the development of immunological assays for the detection of low molecular weight compounds of both natural and synthetic origin. The full potential inherent in the high-affinity antigen-antibody interaction was only realized when labeled antigens of sufficiently high specific activities became available and this resulted in the development of radio- [19], enzyme- [82,350] and fluoroimmunoassays [64,320] for a wide range of haptenic structures. Today, immunoassay is one of the most powerful analytical tools in the biosciences — long established in disciplines such as clinical chemistry and endocrinology, but still rather new to plant analysis. It is the aim of this chapter to review the achievements that have been made in this latter area and to discuss the advantages and limitations of the technique when applied to plant studies.

In addition to pesticides, environmental chemicals and synthetic growth regulators, endogenous low molecular weight constituents, such as natural plant products and plant hormones, will be discussed. Some of the techniques cited in this chapter have not been designed for plant analysis specifically, however, they were included because they have immediate relevance for plant systems. No attempt has been made to incorporate all published material in this review. Rather, the application or applicability of the technique for plant analysis served as a guideline for the selection of the material included. Where appropriate, recent related reviews are cited to facilitate access to other relevant literature. Some emphasis will be placed on the potential offered by the monoclonal antibody technology.

2 Principles of Immunoassay

2.1 General Principles

The following terminology will be used throughout this text: The term *antigen* designates the non-derivatized low molecular weight compound (LMC) which is to be quantitated by the immunological assay. The term *hapten* designates the antigen as coupled to the carrier molecule; this is the antigen (or antigen derivative) plus the linker (or bridge) used in the coupling procedures. The term *immunogen* is used for the hapten-carrier complex used to elicit the immune response; e.g. the hapten bound to a protein carrier. *Epitope* refers to that area of an antigen or immunogen actually bound to a given antibody's antigen combining site. The term *tracer* is used for the labeled antigen or labeled antigen derivative, a the labeled antibody employed to monitor the immunological reaction between antigen and antibody. *Immobilized antigen* refers to a hapten-protein conjugate used in the immobilized-antigen assay mode, i.e. when covalently or non-covalently bound to a solid support. While in the labeled-antigen assays (Fig. 1A), antigen binding to the antibody is directly monitored by the displacement of tracer from the antibody, the other assay format (Fig. 1B) utilizes an indirect detection of specific antibody bound to the immobilized antigen with a labeled second antibody tracer. Both assay formats are the same in that they depend on precise amounts of at least three reactants (1. standard or sample, 2. labeled antigen or immobilized antigen and 3. the specific antibody).

Macromolecular antigens can be analyzed with saturation assays (Fig. 1 C) which employ excess reagents and are less dependent on dosage and antibody affinity than the equilibrium assay discussed above. Since all assays for LMC belonging to this latter type (Fig. 1 a or b), the key prerequisite of all such assays is an antibody or antiserum which exhibits, in addition to high selectivity, high affinity for both the antigen and the tracer.

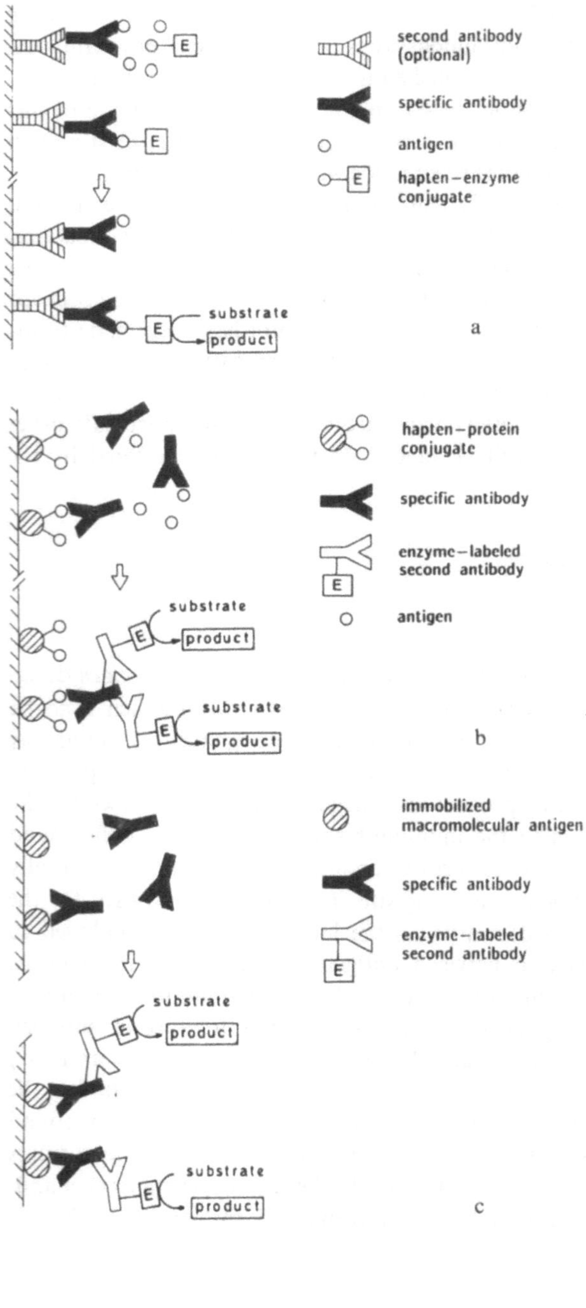

Fig. 1 a-c. Some frequently used solid-phase enzyme immunoassay (ELISA) formats. **a** ELISA for low molecular weight antigens employing immobilized antibody and labeled antigen (direct ELISA); **b** ELISA of low molecular weight antigens employing immobilized antigen-protein conjugates and labeled antibody (indirect ELISA); **c** ELISA of macromolecular, immobilized antigen

A detailed assessment of the mathematical theory of immunoassay is beyond the scope of this chapter. The reader is referred to Refs. [20, 33, 94]. Among the factors of critical importance for assay performance are the quality of the antibody or antiserum and the specific activity of the tracer. Many of the compounds which will be discussed later, but especially pesticide residues or metabolites, environmental chemicals and the plant growth regulators may be present in nanomolar or even lower concentrations. It is evident that only antibodies exhibiting affinities in this range can cope with such samples without the need for extensive concentration. Additionally, the label used must be readily detectable at these concentrations. This requirement limits the choice of labels for most practical applications to radioisotopes, enzymes, and fluorescent or luminescent probes.

2.2 Radioimmunoassay

Despite the existence of non-isotopic immunoassays, radioimmunoassay (RIA) still represents the most frequently used immunoassay, with a calculated number of over 400 million tests performed annually worldwide [76]. An introduction into this technique can be found in [76,100]. The major advantage of RIA over non-isotopic immunoassays lies in the fact that, frequently, the introduction of the label is possible without structurally altering the antigen. Thus, the ideal competitive system with identical structures of tracer and antigen can be realized. Structurally altered LMC may exhibit significantly different antibody-binding than the antigen itself. This can lead to serious problems in assay optimization, especially when polyclonal antisera are being used.

The requirement for high specific activity limits the choice of radioisotopes to the following (spec. activities in TBq/matom in brackets): ^{3}H (1.11), ^{35}S (56), ^{125}I (81), ^{32}P (340), ^{131}I (592), of which ^{3}H and ^{125}I are the most widely used. A certain disadvantage of RIA lies in the fact that it requires a licensed laboratory and rather expensive equipment. On the other hand, fully automated RIA-systems are available which permits the processing of a large number of samples either sequentially or simultaneously (see [112], for a review), a factor which may become important for assays designed for routine screening.

2.3 Enzymeimmunoassay

Whereas radioisotopes are usually employed to label the antigen, immunoassays involving both enzyme labeled antigen and enzyme labeled antibody are common (see Fig. 1). Enzyme immunoassays (EIA) which require separation of the antibody-bound and free fractions of antigen are termed heterogenous EIA whereas assays (e.g. the EMIT assay) which do not require this separation step are called homogenous. Homogenous EIA is usually executed in the liquid phase while the most elegant way of separation of bound and free antigen in heterogenous assays is the introduction of a solid phase to which the antibody or the antigen is bound. Solid-phase EIA is also termed ELISA (enzyme-linked immunosorbent assay). It is in this review not possible to discuss the numerous variations of EIA that have been described. An introduction into this field be found in [257] and more general treatments of EIA in Refs. [219, 337]. Among the enzymes most frequently used in the ELISA are horseradish peroxidase

(EC 1.11.1.7), alkaline phosphatase (EC 3.1.3.1), β-D-galactosidase (EC 3.2.1.23), glucoamylase (EC 3.2.1.3) and glucose oxidase (EC 1.1.3.4). For the analysis of plant material, enzymes which are absent or present in small amounts in plant tissues diminish the risk of assay interference at this stage. Due to its relatively good stability and availability in highly pure form, alkaline phosphatase is a preferable marker enzyme for plant analysis and has been employed in a number of studies, e.g. [14,77,361]. EIA has the advantage over RIA in that its use is not restricted to licensed laboratories and it can be performed with less expensive equipment. On the average, it appears that EIA is also more sensitive than RIA. Some of the published EIA procedures are sensitive down to approximately 5—10 fmol per assay or even lower [159] which is, by a factor of 5—10, better than can be obtained with RIA. When based on appropriately selected monoclonal antibodies (MAB), differences in the affinities of the enzyme labeled and unlabeled antigen can be minimized. This eliminates the major problem in assay development when based on polyclonal mixtures of antibodies.

In RIA, the detectability of labeled antigen is ultimately limited by the specific activity of the radioisotope used and the number of isotopic substitutions introduced into the tracer. The increased sensitivity of the EIA results largely from the signal amplification provided by enzyme catalysis. Sensitivity can be shifted to the extreme if, instead of a colorimetric determination of enzyme activity, fluorogenic substrates [175,310] or radiolabeled substrates [132] are used. With ^{3}H-AMP as substrate and alkaline phosphatase as marker enzyme, Harris and coworkers were able to quantitate as little as 600 molecules of cholera toxin by ELISA [132]. Recently, enzyme amplified ELISA was introduced [160,327]. The assay makes use of an enzyme cascade, in which the product of the primary enzymatic reaction activates a second reaction, or a reaction cycle resulting in an large amplification of the signal [210]. Thus, Johannsson et al. [160] used the conversion of NADP to NAD by alkaline phosphatase as the primary reaction. NAD, in turn, activates a redox cycle involving NAD-NADH cycling catalyzed by alcohol dehydrogenase and diaphorase with the ultimate reduction of *p*-iodotetrazolium violet to an intensely colored formazan dye.

Theoretically, a single molecule of tracer enzyme should be detectable by amplified assays. In practice, the sensitivity which can be achieved in amplified hapten-ELISA (which is necessarily an equilibrium assay), is limited by antibody affinity and by the background generated by unspecific binding, autohydrolysis of substrates and the like.

Recently[403], an elegant technique, enzyme immunochromatography, was devised which permits semi-quantitative to quantitative assays to be done without any instrumentation at all. Although the technique, at present, still lacks the sensitivity of other ELISA techniques, the potential of enzyme immunochromatography for applications such as pesticide screening, or residue analysis, and in field situations, is evident.

2.4 Other Immunoassays

Alternatives to EIA and RIA have been developed such as fluorescent,[64,320] or luminescent,[313,395] immunoassays, spin immunoassay,[203] and electroimmunoassay, [137,280]. Other techniques rely on phage inhibition [7], erythrocyte agglutination. [3], latex bead agglutination,[353], microsphere entrapped dye release, [295], or are based on the changes in light scattering and absorption during fluid-phase immunoprecipitation, [69]. Metalloimmunoassays, [37], make use of organo-metallic labels which can be detected by atomic absorption spectrometry. In addition, differential pulse polarography or anodic stripping voltammetry can be used to detect low levels of haptens labeled with electroactive groups such as mercuric acetate [136]. There seems little limitation to the exploitation of labels other than the imagination of the researcher and the requirement for detectability at low concentrations. In practice, some techniques are still at an experimental stage and sometimes lack sensitivity, are not easily carried out or require expensive equipment. Fluorescent or luminescent labels are of special interest when designing immunoassays for very low levels of haptens. In plant analysis, however, fluorescent labels may be of limited value due to the abundance of autofluorescent LMC which may contribute to background noise. Luminescent immunoassays, on the other hand, do hold great promise for plant analysis, but this reservoir has not yet been tapped.

3 Technical Realization of Immunoassays

3.1 Immunogen Synthesis

Obviously, the way of presentation of the haptenic structure to the immunocompetent cells largely determines the set(s) of cells which will become stimulated and turn into proliferating B-lymphocytes. These subsequently mature into antibody-secreting plasma cells. Thus, immunogen structures become of prime importance, even more so because inevitably, LMC have to undergo structural changes (in the process of linkage to a carrier) before they are used for immunization. Thus, especially for very small molecules which cannot fill an antibody's antigen binding site completely, it is the hapten which is the 'true' antigen and the implications arising from this inference must be kept in mind throughout the development of such assays. Antisera raised against haptens will contain not only antibodies directed at unmodified sites of the antigen but also antibodies directed at the modified structures (e.g. the linkage structure or 'bridge'). This latter antibody population may severely interfere when antigen derivatives which bear similarities to the immunogen are used as assay components. This situation may arise when enzyme labeled haptens are used together with antisera raised against structurally similar hapten-protein conjugates, in EIA. Removal of the bridge-recognizing antibody populations or alterations in hapten structure and/or coupling site(s) may be required in such case. MAB can be selected for absence of such interference and it is strongly recommended that screening procedures which will select for high affinity MAB which bind the antigen proper be utilized.

A second general consideration affecting the design of immunogens relates to the choice of the coupling site. It is generally observed that structures at, or in the vicinity of this site, or which are masked by the bridge, are not or only weakly recognized by the resulting antibodies. Thus, proper choice of antigen derivatization ensures optimum selectivity of the elicited antibodies and will determine the performance of the assay in the experimental situation. The coupling site may strongly affect the ability of the antibodies to discriminate optical isomers, precursors from metabolites, free hormones from conjugates etc. As an example, C_1-couplied cis,trans-(S)-abscisic acid (ABA) gives rise mainly to antisera which recognize cis, trans-(S)-ABA and C_1-derivatives of it, such as the methyl or glucosyl ester [356,380]. C_4'-coupling yields antisera and MAB binding cis,trans-(S)-ABA in its dissociated form with little or no interference from C_1-conjugates [229,381]. Clearly, the latter type of immunogen is better suited for physiological studies on the free hormone. These general precautions aside, there is a wealth of synthetic procedures available to construct immunogenic conjugates for virtually any LMC. Overviews on this area can be found in [85,211]. Proteins such as serum albumins, ovalbumin or hemocyanin are preferentially used as hapten carriers because they are easy to handle, possess a large number of accessible functional groups such as amino-, carboxyl-, or tyrosyl residues, and are highly antigenic themselves. It may not be necessary to optimize the hapten-to-carrier ratio. In several cases, a strong immune response was elicited in rabbits and mice against various hapten-protein conjugates substituted in a mere 1:1 ratio (unpublished results from our laboratory). The majority of hapten-protein conjugates synthesized in our laboratory turned out to be sufficiently immunogenic when the coupling ratio ranged from 5—10.

3.2 Antibody Production

The production of the key element of all immunoassays, the antigen-specific antibody, requires special consideration with respect to the choice of animals to be immunized, the immunization routes, collection, storage and further processing of the immunoglobulins. These factors shall be considered now, with special emphasis given to the generation of MAB.

3.2.1 Immunization

Detailed immunization schedules have been published for a number of mammals. Most laboratories will perhaps focus on rabbits for the production of antisera and on mice or rats for the production of MAB.

For rabbits, combined foot-pad and intramuscular immunization, e.g. [372], multiple-site intradermal immunization, e.g. [252] and, most commonly, combined intradermal and intramuscular routes, e.g. [361] are employed. A standard protocol which gives consistent results with a large number of hapten-albumin conjugates can be found in [361]. However, many other protocols have been published and used successfully. Multiple-site immunizations and foot-pad preimmunization may give higher antibody titers and a better response with weak immunogens and should be considered when routine protocols give unsatisfactory results. The administration of 50–200 µg per animal per injection of serum albumin-based immunogens emulsified in adjuvant (e.g. complete Freund's), or adsorbed onto charcoal, talcum powder or similar supports [100] will stimulate a sufficient primary response. Booster injections usually require less immunogen and it is frequently possible to omit the adjuvant or to use incomplete adjuvants. Overdosing of immunogen might result in immunotolerance which will result in falling antibody titers despite continued immunization.

Rodents are usually immunized via an intraperitoneal or a combined intraperitoneal/intravenous [113] route. The amount of immunogen is largely dependent on the protein carrier used. Hemocyanin, which is an extremely potent antigen *per se*, must be used carefully and in low quantities (a few micrograms or less per injection). Serum albumin is also very antigenic, but hapten-albumin conjugates can be used in the wide range from 10–200 μg per injection (intraperitoneal). General immunization schedules have been published, e.g. [361] but a deviation from these may be required in special cases. Both short-term and long-term immunization schedules are used for rodents with the aim of MAB production. In our experience with LMC specific MAB, long term immunizations performed as described [361] gave higher rates of positive hybridomas (unpublished results) than short term protocols. There are exceptions however, and it is adviseable to use a flexible schedule for new types of haptens.

In vitro immunization [292] is receiving increasing attention because it can be carried out with very small amounts of immunogen under defined conditions and may be completed in less than 3–4 days, thus speeding up the whole MAB process considerably. The technique has not been used extensively for haptens, but there are reports that *in vitro* immunization gives hybridoma yields comparable to the *in vivo* process [32]. It can be expected that this elegant technique will become increasingly important.

3.2.2 Polyclonal Antibodies (Antisera)

Immunization of rabbits can be continued for 12–18 months with an expected yield of 450–500 ml serum over this period. Depending on the titers, this amounts to 10^5 to 10^7 RIA or ELISA analyses which can be run with serum from a single animal. This yield will satisfy even the largest needs and thus antiserum production will be economic for most purposes. It is important to note, however, that individual serum collections from the same animal, and even more so, from different animals, will show considerable variations as to titer and specificity. Consequently, serum pools should be established, but only after a preliminary screening of the individual serum samples. Long term storage of pooled sera can be done aliquotted in the frozen state or as a lyophilized, 50% saturated ammonium sulphate fraction, after removal of the salt. The stability of frozen sera extends over many years and even decades. Repeated freezing and thawing, however, should be avoided.

It may be necessary to purify sera further prior to use, e.g. to remove interfering antibodies or to concentrate specific ones. A variety of techniques for this purpose is available including ion exchange or affinity chromatography on immobilized protein A-, double antibody-, or hapten-columns [40,158,268,326]. Likewise, immobilized carrier-plus-bridge can be used to deplete sera of unwanted antibodies [178a]. These steps can be highly efficient but are nearly inevitably accompanied by substantial losses of the desired antibody.

3.2.3 Monoclonal Antibodies

Detailed laboratory manuals describing all steps of MAB production have been published, e.g., [38,113,279], thus the discussion herein can be restricted to some general remarks pertinent to the production of LMC-specific MAB suitable for immunoassay.

It is frequently argued that a MAB based system cannot be as specific as polyclonal antibody based ones and this appears to be the case with large, multiple-epitope antigens. Here, antigen precipitation results from a network of reactions involving bridging, and crosslinking of different epitopes. A single antibody, such as a given MAB, recognizes only a single etipope, thus reducing the overall selectivity of the process. MAB usually do not precipitate their macromolecular antigens unless the epitope recognized is present more than once on the same macromolecule.

The situation is different with LMC especially in immunoassay systems. LMC usually consist of a single, or a very limited number of epitopes and the antigen-antibody binding process as observed in immunoassays is (because of the low levels of tracer-antigen) usually dominated by one or only a few populations of antibodies (those with highest affinities). This results in a 'quasi-monoclonal' behaviour of antisera at least at low antigen levels. It is thus not unexpected that comparisons made between MAB- and antisera-based immunoassays have so far shown the MAB variant to have equal or similar specificity [229,381].

The basis of antibody diversity has recently been unraveled to a large extent, and it is now clear that an individual mouse is able to produce approximately 2.4×10^4 $V_\varkappa$ variants (about 95% of murine immunoglobin light chains belong to the $\varkappa$ type) and about 9.6×10^5 V_H variants [177]. Given the random combination of these V-regions, an enormous number of 2×10^{10} different antigen-binding sites can be created [177]. Earlier studies have shown [177] that in the range of 10,000 different antibodies can be directed against a single haptenic structure, such as the *o*-nitro-*p*-iodophenyl-moiety. It is this incredible potential of the immune system which one taps when producing a MAB against LMC, a potential which has not been even partially realized to date. Thus, the crucial importance of a proper screening procedure to select the wanted MAB becomes clear. MAB selection is to some extent a statistical problem and, even with the best of all screening, requires patience and a good deal of luck.

Immunoassays are frequently developed with the aim of increasing analytical sensitivity. In this case, the requirement for high antibody affinity is stringent. Affinity constants should be in the range of 10^8–10^{10} M^{-1}. It is evident that assays employed to screen for hybridomas secreting such antibodies, must be suitable to detect antibody or antigen binding at nanomolar levels. A second stringent conditions is that not only the haptenic structure, but also the unmodified LMC must bind to the antibody with equal or at least similar affinity. This must be taken into consideration when designing the screening assay. In our experience, the only assay which combines both requirements in a single step is RIA using the unmodified antigen as tracer. This strategy has permitted the selection of a set of high-affinity MAB specific for plant hormones with almost no false positives [77,79,229,230]. Screening by ELISA initially identifies more positive hybridomas, but this is due to the detection of low affinity MAB (especially in the immobilized-antigen mode, Fig. 1) and bridge-recognizing antibodies. Reversibility of biding by the unmodified LMC must then be shown and this duplicates the number of screening assays.

Hybridoma cultures secrete MAB mainly near and during the stationary phase of their growth cycle. Supernatants from stationary phase cultures may contain between 10 and 100 mg l^{-1} of MAB. From such culture supernatants, MAB can be fractionated by salt precipitation, ion exchange chromatography, HPLC and some subclasses also by protein A affinity chromatography (see: [32,38,40,113,158,268,279]). Provided that serum free media have been used, highly pure MAB will be obtained. In serum-containing media, considerable contamination with foreign immunoglobulins will result which can only be removed by affinity chromatography. Both procedures are tedious and costly on a preparative scale. A cheap alternative to this is ascites production. Ascites fluid from the intraperitoneal cavity of mice inoculated with hybridoma clones may contain as much as 10–20 mg ml^{-1} of MAB, however, these will also be contaminated by immunoglobulins and other proteins from the animal. Thus the preparation of pure MAB from ascites also requires affinity purification steps.

If large quantities of MAB are needed, more sophisticated techniques may be employed, which yield highly concentrated solutions of MAB uncontaminated by other immunoglobulins. An array of such processes suitable for industrial scale-up is available including hollow-fibre culture [6], dialysis fermentation [2] and cell micro-encapsulation [74]. If the MAB is desired solely for immunoassay purposes, however, pure antibody is not necessary and dilute ascites or immunoglobulin fractions obtained from cell culture supernatants by salt fractionation are perfectly suitable. The contaminating proteins can act as 'protectants' by reducing unspecific adsorption of MAB to solid supports, or as 'carriers' facilitating complete precipitation of antibody in many RIA procedures.

3.3 Assay Protocols

It is evident that in order to maximize precision, assay protocols need to be designed which involve a minimum number of pipetting steps and minimum sample manipulations. For large series, automated equipment can be considered and this is available from several commercial sources. In spite of the diversity of potential applications, general assay protocols cannot be given. Examples of protocols optimized for plant analysis by RIA and EIA can be found in refs. [361,375].

3.4 Assay Validation

Quantitative analysis is a statistical rather than absolute phenomenon, and two fundamental and independent types of error can be distinguished: random and non-random. Random error determines the precision and non-random error determines the accurary of the analysis [297]. Precision is defined as the concordance between a series of measurements of the same sample and accurary as the degree of approximation of the 'true' value by the analysis.

Obviously, checks for assay performance must include an assessment of precision as well as of accuracy and the evaluation programs used must reflect the sources of error inherent in immunoassay.

Random error is largely determined by factors which also operate in other analytical systems (e.g. operator skill, precision of pipetting devices, uncertainties in fitting standard curve calibration points etc.) and will not be discussed here. A detailed treatment is given in [222,267]. Immunoassay precision typically ranges from 10–20%

(coefficients of variation for replicate analysis run in different assays) but can be much higher especially when antigen levels have to be determined from the lower end of standard curves.

Some attention shall be given here to non-random error affecting accuracy because these factors tend to be more assay specific and may not be obvious. Table 1 gives some of the major factors which might contribute to non-random error in immunoassay (see also refs. [267, 361, 375]).

Table 1. Some sources of non-random (systematic) error in immunological analysis

- Presence of cross-reactants
- Presence of factors destabilizing or desorbing antibodies (e.g. surfactants, phenols)
- Presence of factors influencing the antigen-antibody interaction
- Antigen binding by other assay components (surfaces of vessels, carrier proteins such as albumins etc.)
- Presence of factors influencing the activity of the marker enzymes (e.g. inhibitors) in ELISA, sample autofluorescence (in FIA), radioactivity quenchers in RIA.

The quality control procedures available to monitor immunoassay reliability have been the subject of several recent reviews [43, 73, 222, 267] and shall be touched only briefly. They include checks for intra- and inter-assay precision. Sample dilution analysis in the presence of calibration standards reveals cross-reactants or other types of interference resulting in non-linearity and/or non-parallelism of sample dilution and standard curves. Furthermore, the presence or absence of cross-reactants can be monitored on immunohistograms and/or in antigen containing fractions subjected to a sequence of successive purification steps (successive approximation, [297]). Recovery is best assessed by the inclusion of radioactive internal standards or spiked split-extracts and a final check for accuracy, whenever possible, should include an inter-laboratory study, applying different analytical techniques to the same set of samples. Due to the complexity of standard curves, it is also recommended that one uses statistically valid data processing procedures rather than rely on subjective evaluation of calibration curves and data points. Guidelines for immunoassay data processing have been presented [73] and software for immunoassay data processing on most common microcomputers is commercially available.

4 Immunoassays for Plant Secondary Metabolites

4.1 Scope

Immunoassay was used in a range of studies (see Chapts. 4.2 and 4.3) to quantitative secondary metabolites in plant tissues. The main scope of these studies was the screening for metabolite formation and distribution in differentiated plants ([251, 369, 370, 372]) and/or in cell cultures (e.g. [106, 122, 123, 384]), and immobilized cells (e.g. [30]), early interest was focused on pharmaceutically important constituents. Immunoassays have been used to obtain surveys of the taxonomic distribution of certain steroidal compounds [364] or alkaloids [389] with the aim of establishing better yields or find as yet unexploited plant sources for these natural products. In addition, an immunoassay was found useful for monitoring biotransformation reactions [377] and the technique has been employed as sensitive and specific assay for enzyme activity [339]. Thus, immunoassays have been used in basic as well as applied studies on higher plants, making use of the high degree of selectivity inherent in the assay and the fact that very large numbers of samples, up to more than a thousand per day per worker, can be processed simultaneously. Immunoassay should prove useful in genetic engineering as well as biotechnology programs which involve the selection for high product yields from differentiated plants, plant cell cultures or microbial strains, the screening for product formation in transformed organisms and during strain optimization and/or selection. Furthermore, the monitoring of fermentation and biotransformation processes with the possibility of direct discrimination between optical isomers, e.g. [229], the monitoring of enzymes at low levels when other assays do not exist or are too insensitive are additional areas which should benefit from use of immunoassays. With respect to the progressing extinction of species and the shrinking of gene resources, the potential of surveying a very large number of species for rare natural products or groups of products (with use of group-selective antibodies) becomes of special importance. Given the sample capacity of ELISA techniques, virtually all the known species of higher plants could be screened for any given metabolite in less than 2 years. For such an analysis, milligram or sub-milligram samples of herbarium specimens could be used in most cases [364].

4.2 Secondary Metabolites of Higher Plants

4.2.1 Steroids and Related Compounds

The assays discussed here were all designed for the pharmacological analysis of plant derived pharmaceuticals or for plant analysis directly.

Among the most important pharmaceuticals from plants are the cardiac-glycosides, strong inhibitors of the Na^+/K^+-ATPase involved in nerve and muscle function. The positive inotropic action of heart glycosides forms the basis of therapy of congestive heart failure, one of the most widespread diseases of western civilizations. Heart glycosides have a narrow therapeutic concentration range and thus require precise dosage and monitoring of drug levels when signs of intoxication become suspect. Immunoassays for *Digitalis* glycosides have thus been developed rather early [34,319] and are being used extensively in the monitoring of clinical *Digitalis* treatment.

Periodate oxidation of the terminal digitoxose of digitoxin (*1*) and digoxin (*2*) facilitated the synthesis of immunogenic protein conjugates [319,372], the production of antisera [251,319,372] and, recently, monoclonal antibodies [32,52,157,281] specific for the aglycone structure. While such antibodies discriminate the various aglycone structures, they tolerate a range of glycosidic residues found at C-3 of the aglycone. On the other hand, immunogen synthesis via attachement of the aglycone to the protein through the lactone ring [336] or through C-17 or C-12 [316] gave rise to antisera sharply discriminating digoxin (*2*) from the bis- and monodigitoxide or the aglycone. This is a classical example of the importance of an appropriate immunogen- and hapten-structure to elicit antibodies with the desired properties. Specific antisera against digitoxin- (*1*) [106,259], digoxin- (*2*) [34,319,372] and gitoxin- (*3*) type [Gaas, W., Weiler, E. W., unpublished] cardiac glycosides, as well as against strophantin [Weiler, E. W., unpublished] and 16-acetylgitoxin (*4*) [365] were reported and apart from their frequent use in the clinical field [35], several were applied to plant studies. Radioimmunoassay for digitoxin- (*1*) and expecially digoxin- (*2*) related glycosides proved sensitive enough to detect between 0.1 and 1 ng of the compounds, and were employed in several studies to determine cardenolide distribution in plants [351,372,376], to select high yielding individuals for breeding programs [351,370,376], to monitor 12β-hydroxylation by *Digitalis* cell suspension cultures [376,377] and immobilized *Digitalis* cells [30], or to study cardenolide production in tissue culture including regeneration studies [106,122,123,124,125].

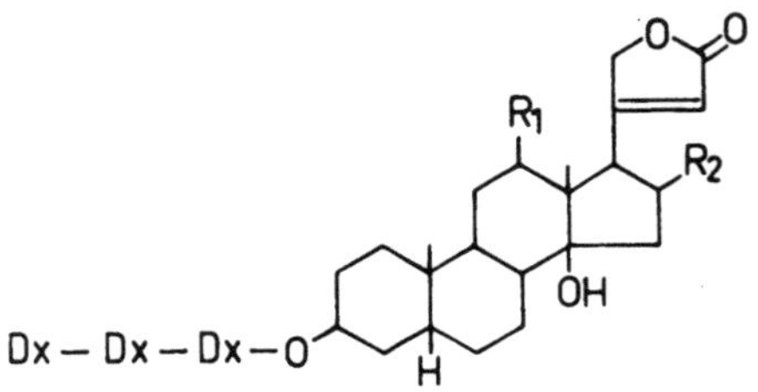

(1) R_1 = H R_2 = H

(2) R_1 = OH R_2 = H

(3) R_1 = H R_2 = OH

(4) R_1 = H R_2 = $OCOCH_3$

Periodate oxidation was used to functionalize the steroidal glycoalkaloid, solasodine (*7*), before coupling to BSA [364]. This facilitated the production of antisera which specifically recognized the spirosolanoles of the solasonine (*5*) and tomatidine series with no interference from solanidanes, spirostanols or phytosterols [364]. In RIA, ng amounts of solasodine (*7*) a potential diosgenin-substitute for the production of semi-synthetic steroids can be quantitated and this assay was used in the screening of *Solanum laciniatum* and *S. aviculare* for high-yield individuals as well as for a phytochemical survey of the genus *Solanum* [364]. Vallejo and Ercegovich [344] coupled solanidine through its 3-OH group to BSA and obtained antisera which recognized the aglycone, but not the 3-glycosidic forms present in plant material. This necessitated a lengthy hydrolysis of plant extracts prior to analysis and sacrified some of the advantages of the immunological assay. The coupling of periodate oxidized α-solanine to BSA allowed the production of antisera recognizing a range of potato glycoalkaloids, e.g. α-solanine, α-chaconine and solanidine [227,242]. This resulted in assays (RIA, ELISA) for total potato glycoalkaloids in tubers [237,241,242] which were comparable to chemical methods. In addition, the assay allowed a correlation of the dietary intake of potato to the serum solanidine levels [227]. This alkaloid was detected for the first time in human serum.

Fong et al. [98] have reported on a radioimmunoassay for the anti-tumor secotriterpene, bruceantin. This hapten was linked to BSA as the hemisuccinate, however, the specificity of the assay which is sensitive in the low ng range, is not known and no application to plant material was reported.

Other steroidal compounds found to be immunogenic include the classical substrate for Marker-degradation, diosgenin (*8*) [308], one of the active constituents of the oriental drug, ginseng, ginsenoside RG_1 (*9*) [307], glycyrrhizin (*11*), and its aglycon, glycyrrhetic acid, a principal constituent of the oriental remedy, *Glycyrrhizae Radix* [169,171,171a], the triterpenoid saponin-glycoside mixture, aescine, [199], the intensely bitter seco-triterpene, quassin (*12*), found in members of the tropical family *Simaroubaceae* which has uses as a food additive and insecticide [298,299], the insect moulting hormone, β-ecdysone, also a constituent from plants (especially found in rhizomes of

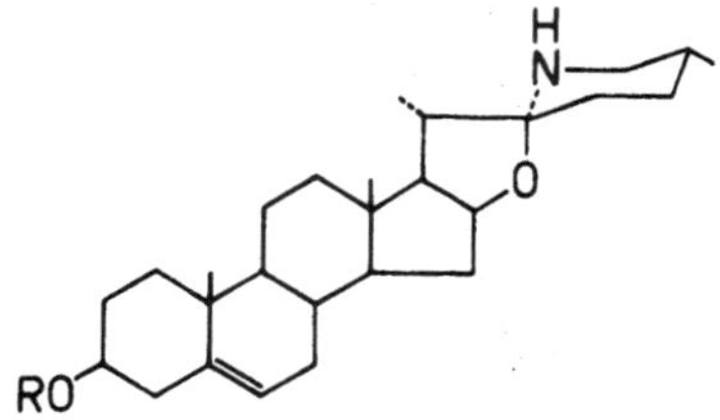

(5) R = –Gal–Glc
Rham

(6) R = –Gal–Rham
Rham

(7) R = –H

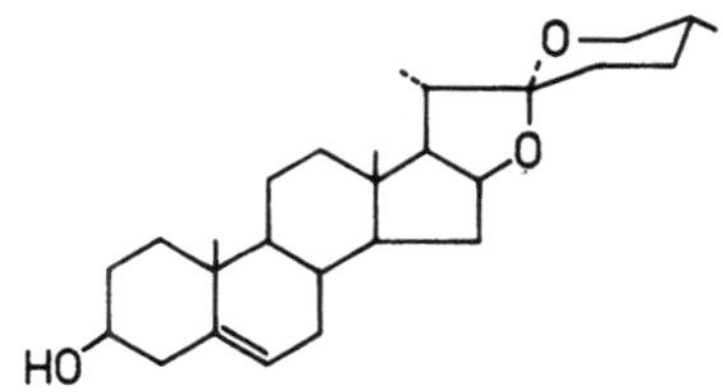

(8)

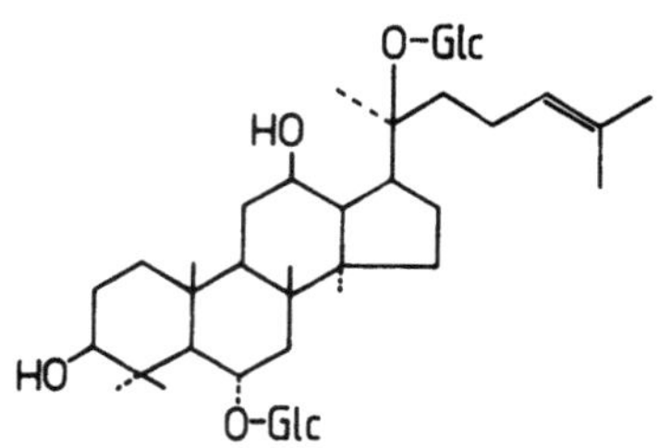

(9)

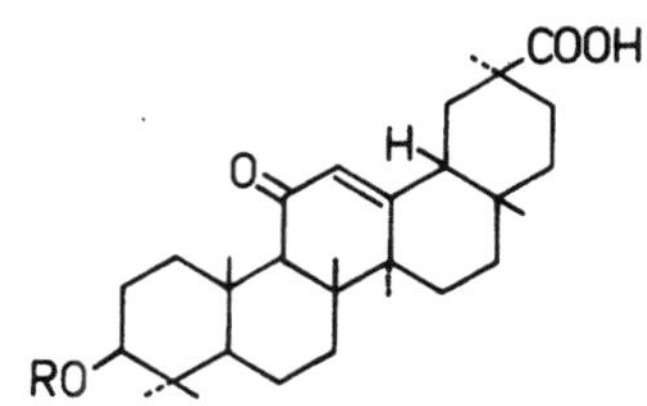

(10) R = H

(11) R = Glcr–Glcr

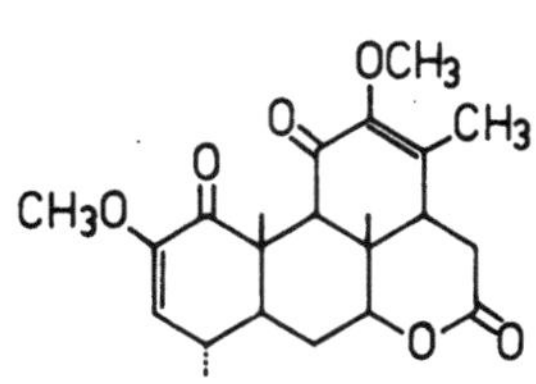

(12)

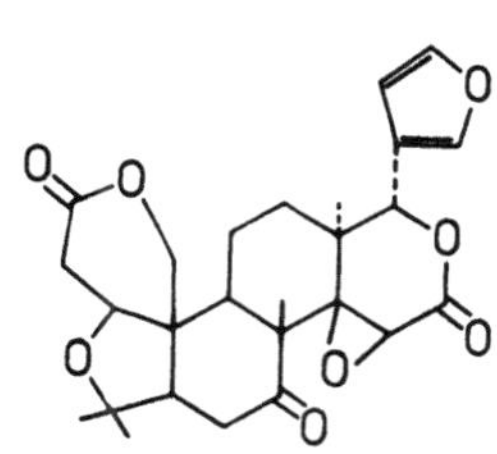

(13)

some ferns [68,284]), the saponin, digitonin [30a], and finally the triterpene-dilactone, limonin (*13*), which is the principal bitter constituent of grapefruit [225,366]. The latter assay has been applied extensively to plant analysis [163,224,225,228,366]. Protein coupling of terpenoids was effected through 0-carboxymethyloximes [225,284,366], hemisuccinates [298,308], through activated esters of carboxylic acids [169,171,199] or through carboxylate azides [307]. Especially sensitive analysis (pg range) was achieved using ELISA [163,225,298,299,366]. Immunoanalysis was employed to study limonin (*13*) distribution in *Citrus paradisi* [223] and has furthermore proven useful for the analysis of the bitter principle in grapefruit juice [228] and callus cultures [16a].

Two other plant constituents, partly of terpenoid origin, were found to be immunogenic as haptens, namely paeoniflorine (*14*) (through 6′-hemisuccinyl and 6′-hemiglutaryl-protein conjugates), a constituent of the oriental medicine "shakuyaku" (Chinese paeony root)[170] and the marijuana hallucinogen, Δ9-tetrahydrocannabinol (*15*) (see review[301] and refs. therein). Furthermore, a commercial assay was used to detect and estimate the content of the secosterol, 1α, 25-dihydroxy-cholecalciferol, vitamin D (*15a*), and its derivatives in calcinogenic plants such as *Solanum glaucophyllum* and *Cestrum diurnum*[383a]. Kolossa et al. have successfully raised antisera against the diterpenoid,α-, and β-2,7,11-cembratriene-4,6-diol (*15b*). a putative flavour precursor in *Nicotiana tabacum* leaves[180b]. The 6-0-succinyl derivative of α-2,7, 11-cembratriene-4,6-diol (*15b*) was coupled to BSA to obtain the immunogen. A very detailed assessment of assay properties is provided as well as an analysis of cembratrienediol distribution in *N. tabacum* plants and leaf tissues, down to the cellular level.

It is thus clear that plant terpenoids, when coupled to suitable carriers, are highly immunogenic and elicit the production of specific, high-affinity antibodies suitable for immunoassay analysis. The applicability of such assays for plant analysis has been proven in quite a few cases. The first immunoassay designed specifically for plant analysis which became commercially available was an ELISA[40a,163] developed for the quantitation of limonin (*13*).

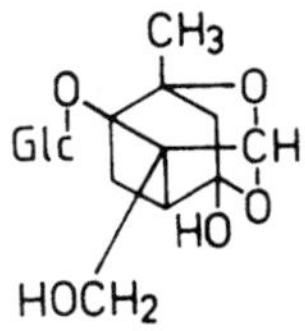

(14)

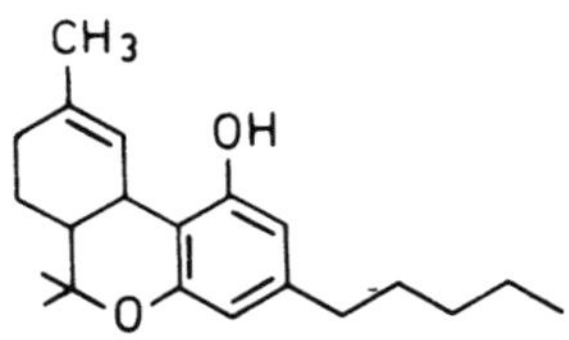

(15)

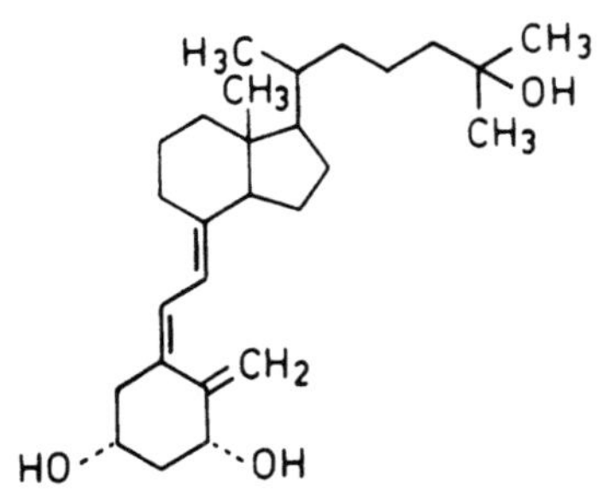

(15a)

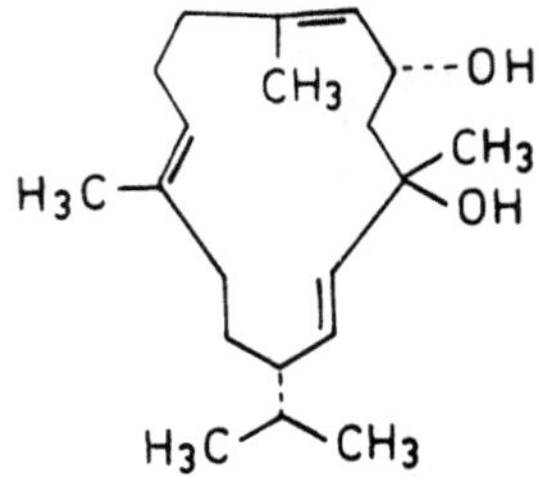

(15b)

4.2.2 Alkaloids

It is not surprising that the immunological assay of alkaloids has attraced much attention, given the pharmacologic importance of these compounds. Unfortunately, they usually occur at rather low levels and are often complex structures which require sophisticated analytical tools for quantitative analysis. The basis for assay development and utilization of these assays is much the same as mentioned in Chapt. 3.2.1 for the steroids. Most reported assays were designed for indole alkaloids and tropane alkaloids, however a wide range of immunoassays for other alkaloids is available as well. Most assays were developed with rabbit antisera, but monoclonal antibodies against certain alkaloids have been described recently [22,22a,283].

4.2.2.1 Indole Alkaloids

The main interest in developing immunoassays against indole alkaloids is either for the monitoring of therapy or phytochemical analysis. In addition, biosynthetic studies have benefitted from the use of sensitive, product-selective immunoassays [339]. We shall proceed in our discussion from assays for certain alkaloid precursors to a discussion of the monomeric indole alkaloids, then the dimeric indole alkaloids and, finally, the highly complex peptidyl indole alkaloids.

Precursors

Antibodies of high affinity ($K_a = 1.6 \times 10^9$ l/mol) against the monoterpene glucoside precursor of a multitude of indole as well as ipecac alkaloids, loganin (*16*), were raised [332] though 6′-carboxyloganin coupled to BSA. These were highly selective and they discriminated loganin (*16*) from secologanin, which is a direct precursor of indole alkaloid biosynthesis. The radioimmunoassay described was sensitive (0.5 ng loganin (*16*) detectable per assay) and was used in conjugation with a second assay, specific for secologanin, to follow the uptake and metabolism of loganin (*16*) in *Lonicera* and *Weigelia* tissue cultures. Likewise, RIA was used by Treimer and Zenk [339] to study, *in vitro*, the fate of secologanin by following the condensation of tryptamine and secologanin with an end-product specific (ajmalicine (*18*)) assay. The ability to discriminate between substrates and products of enzymatic reactions and the sensitivity of immunological assays (which is matched only by radiotracer assays) make such assays an attractive choice for *in vivo* and *in vitro* enzymatic studies even for low-level enzymatic activities.

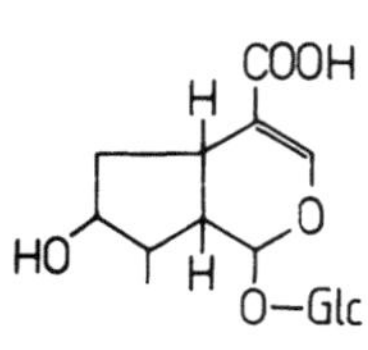

(16)

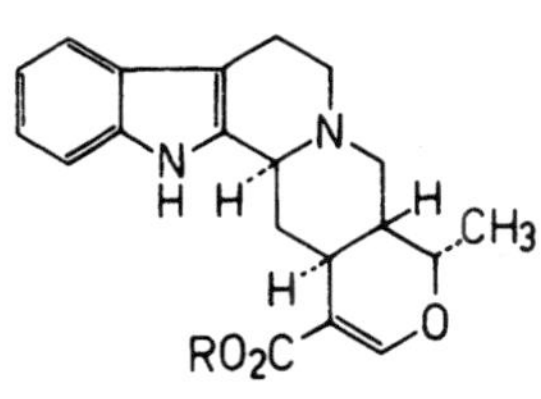

(17) R = H
(18) R = CH3

Monomeric Indole Alkaloids

Immunoassays have been developed for the following monomeric indole alkaloids: ajmaline [10a], ajmalicine (*18*) [8], serpentine (*20*) [8] and reserpine (*210*) [205] which are important because of their use in the treatment of circulatory diseases such as hypertension, and catharanthine [68a] and vindoline (*22*) [182,191a,384], which are important because of their suitability as monomers for the synthesis of the antileucemic dimers of the vinblastine (*24*) type.

Both ajmalicine (*18*) and serpentine (*20*) were hydrolyzed to the corresponding acids (*17*, *19*) which were activated with N-hydroxisuccinimide, then coupled to human serum albumin (HSA)[8]. For reserpine (*21*) a different approach was followed. This compound was protein-linked through the indole-N (Mannich-reaction) or through the indolyl-benzene-nucleus via diazo-coupling[205]. Vindoline (*22*) was first deacetylated to yield deacetylvindoline and this was converted to deacetyl-vindoline-17-O-glutarate[384] or desacetylvindoline-17-O-hemisuccinate[182] before being coupled to protein. All four coupling sites proved to yield effective immunogens and these gave rise to selective, high-affinity antibodies, which allowed the RIA detection of ng quantities of the alkaloids. To date, the assays for serpentine (*20*), ajmaline[10a], ajmalicine (*18*)[8], catharanthine[68a] and vindoline (*22*)[191a,384] have been applied to plant analysis.

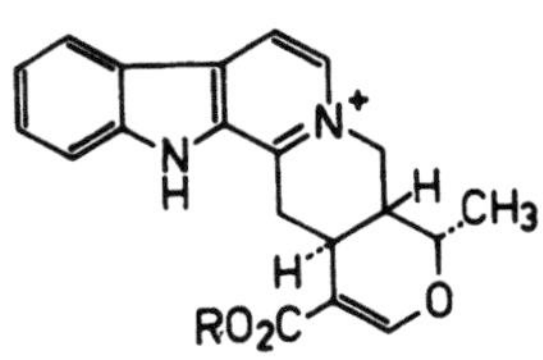

(19) R = H
(20) R = CH_3

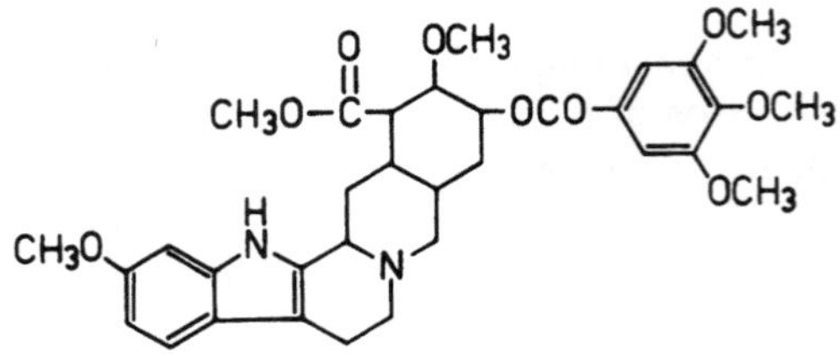

(21)

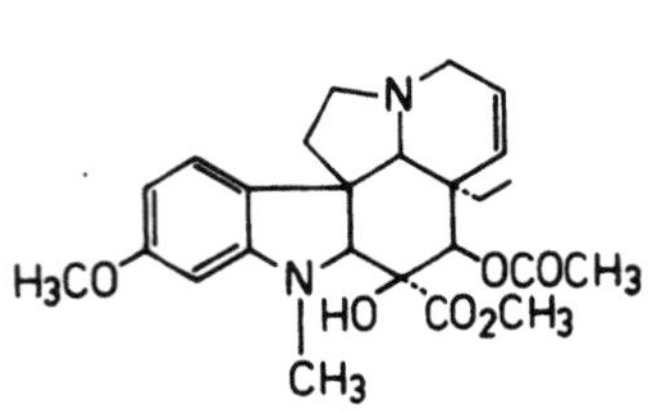

(22)

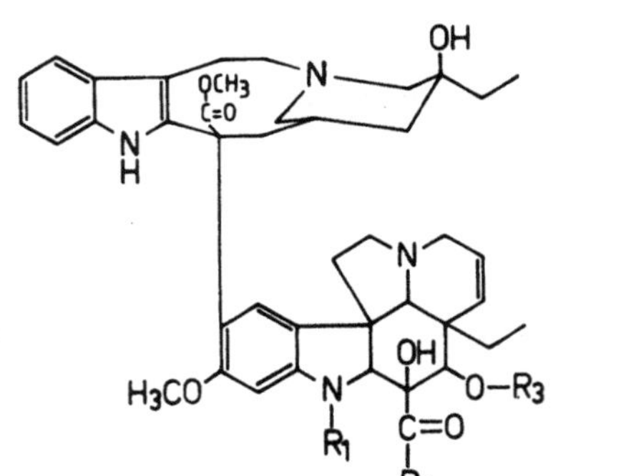

	R_1	R_2	R_3
(23)	CHO	OCH_3	$COCH_3$
(24)	CH_3	OCH_3	$COCH_3$

Dimeric Indole Alkaloids

The dimeric indole alkaloids, vinblastine (*24*) and vincristine (*23*) have considerable therapeutic value because of their pronounced antineoplastic properties, especially in the treatment of leucemic children. Both compounds have to be analyzed at tissue levels in treated patients and occur at low abundances in the plant source (*Catharanthus roseus*). It is not surprising, therefore, that a range of assays, e.g. [153 d, 187, 191, 283, 302, 335] was developed for both, vinblastine (*24*) and vincristine (*23*); most of them for clinical application. While antibodies raised against Mannich-coupled vinblastine (*24*) [335] showed little discrimination of either alkaloid, linkage through an open ring dicarboxylic intermediate prepared by periodate/permanganate oxidation of the olefine 6,7-double bound of both, vinblastine (*24*) and vincristine (*23*) resulted in much better (ca. 200-fold) discrimination [187]. Most workers have used the 4-deacetyl-3-hydrazide intermediates of the alkaloids for immunogen synthesis which gives little discrimination between vinblastine (*24*) and vincristine (*23*) [314], however, more specific monoclonal antibodies have recently been developed [283]. A high selectivity was obtained linking vincristine (*23*), through the 12′-position of its catharanthine moiety, to diazotized aromatic carboxylic acids, such as *p*-aminobenzoic acid. The resulting derivative was in turn reacted with protein-aminogroups [191]. Due to their high antibody affinity, the resulting antisera allowed the quantification of vincristine (*23*), in the low pg range, by ELISA [191]. Little work has been done applying these assays to the phytochemical analysis of the tumor alkaloids in *Catharanthus*, except for preliminary experiments [187, 191] which also revealed interference from other sources, possibly monomeric precursors, when crude extracts were analyzed [191].

Claviceps Alkaloids

Antibodies and immunoassays against lysergic acid (*26*) derivatives, such as the hallucinogen, LSD, have been described, for clinical and pharmacological analysis, e.g. [42, 178, 209, 333, 349]. Arens und Zenk [10] have reported the synthesis of different lysergic acid (*26*) immunogen structures allowing them to raise either polyspecific antisera, reacting with lysergic acid (*26*) and a range of simple carboxylic derivatives, e.g. (*25*), or monospecific antisera, reacting exclusively with lysergic acid (*26*) [284]. This allowed them to determine the compound in sclerotia and saprophytic cultures of the fungus, *Claviceps purpurea*, and to optimize fungal cultures for the production of ergometrin. These authors have extended their studies to the immunological assay of the peptidyl ergot alkaloids, ergotamin (*27*), and ergocristin (*28*) [9] which have considerable therapeutic value and are industrially produced largely from parasitic

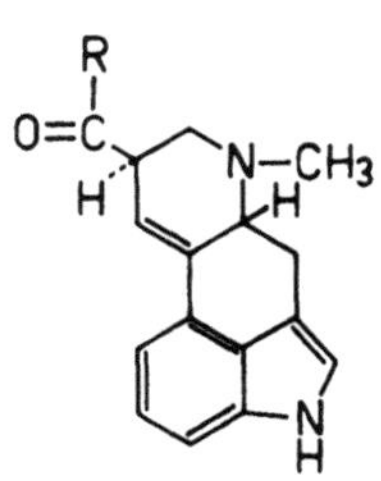

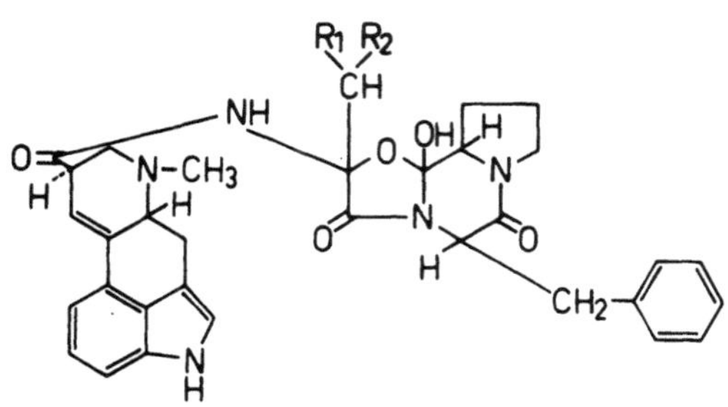

(25) 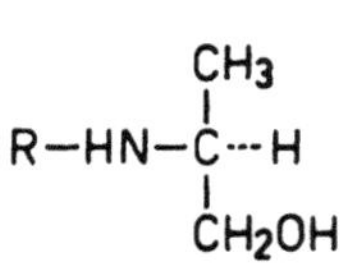

(26) R–OH

(27)	R1 = H	R2 = H
(28)	R1 = CH_3	R2 = CH_3

cultures of *Claviceps*. The assays [9] were designed to help facilitate the optimization of saprophytic fungal cultures and fermentation processes for alkaloid production, are sensitive down to the ng-level and were applied to ergotamin (*27*) and ergocristin (*28*) analysis in *Claviceps purpurea*. Immunoassays for the clinical investigation of ergot alkaloids were reported later, e.g. [303,369].

4.2.2.2 Tropane Alkaloids

As before, interest in the immunological assay of tropane alkaloids stems from the pharmacological importance of the major compounds, hyoscyamine (*29*) and, most notably, scopolamine (*30*). These alkaloids are produced from plant material (usually leaves) of *Datura* or *Duboisia* species, and considerable effort is being made towards improving the plant's productivity as well as attempting alkaloid production in tissue culture. Weiler et al. reported on the synthesis, via nor(—)scopolamine-N-β-propionic acid, of immunogens which resulted in antisera showing very little interference from hyoscyamine (*29*) [369]. These antisera have been used in an RIA to quantitate scopolamine (*30*) directly in unpurified, dilute, leaf extracts of *Datura sanguinea* and *Duboisia* hybrids [369]. This assay, and a more sensitive ELISA based on the same serum (Zenk, unpublished), was further applied to the analysis of somaclonal variation of scopolamine (*30*) content in cell culture clones of *Hyoscyamus muticus* [258]. A radioimmunoassay for atropine ((±)hyoscyamine (*29*)) [200] was reported to be applicable to the direct analysis of plant extracts, after base catalyzed racemization of the endogenous (—) hyoscyamine (*29*). Recently, anti-tropic acid antibodies were obtained immunizing rabbits with tropic acid-BSA conjugates [97 a)].

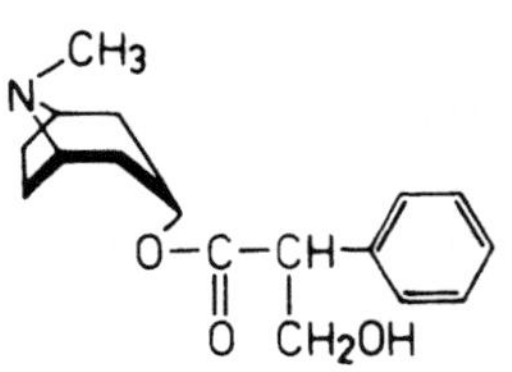

(29)

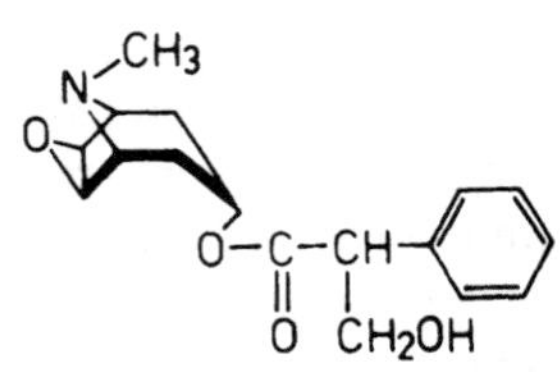

(30)

4.2.2.3 Other Alkaloids

Immunoassays have been reported for a variety of other alkaloids, mainly of clinical importance. Among those of relevance for plant analysis are the *Nicotiana* alkaloids of the nicotine-type, the *Papaver* alkaloids of the morphine type, certain *Cinchona* alkaloids, colchicine and the methylxanthine, theophylline.

Nicotiana Alkaloids

The reported assays are primarily concerned with the monitoring of nicotine (*31*), its major metabolite in humans, cotinine or N′-nitrosonornicotine (*32*), a precarcinogen suspected of contributing to the tumorigenic activities associated with tobacco smoke, in active and passive smokers. The functionalization of nicotine (*31*) was effected in a number of ways, e.g. through the aminoethylpyridinium derivative [255], 4′-hydroxymethyl-nicotine hemisuccinate [189], 6-aminonicotine [127], 6-(*p*-aminobenzamido)nicotine [226], N′-nitroso-5′-carbonylnicotine [41] or, most recently, though 3′-hydroxymethyl- and 4′-carboxynicotine as well as — cotinin [22]. In spite of the limited sensitivity, (thus precluding the monitoring of nicotine (*31*) and cotinine, e.g., in passive smokers), the elegant work of Langone's group [22] once again demonstrated the tremendous potential of the monoclonal antibody technology. These workers have selected high-affinity MAB perfectly tailored to operate in high-performance ELISA systems which were able to measure as little as 50–100 femtogram of nicotine (*31*) or cotinine. Except for an unpublished study (Weiler and Zenk, unpublished) showing the feasibility of immunoassay for large-scale screening of dihaploid tobacco plants for low and high nicotine (*31*) individuals, these assays have not been adopted for plant analysis, although in terms of sensitivity and selectivity [22], existing assays should be well suited for this task. Both polyclonal and monoclonal antibody-based immunoassays for cotinine were reported, e.g. [22 a, 178 a].

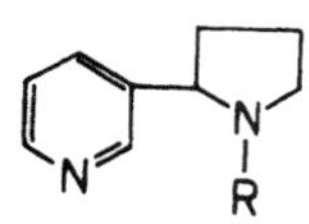

(31) R = CH_3

(32) R = NO

Papaver Alkaloids

For obvious reasons, the necessity of clinical or forensic monitoring of morphine (*33*) and codeine (*34*) levels has triggered a relatively early development [324] of immunological assays for these addictive drugs. The major emphasis was on assays with low limits of detection and short analysis times. In addition, it was important to discriminate between the two closely related structures. To this end, N-carboxypropyl-normorphine [96], N-butyroyl-norcodeine [97] or N-carboxyethyl-normorphine [111], or N-carboxybutyl-normorphine [51 b] have proven suitably functionalized derivatives resulting in antisera with minimal cross-reactivities against the other alkaloids. Other synthetic approaches included the use of 6-succinylmorphine [126] or 3-O-carboxy-methylmorphine [325].

There are but few reports on the phytochemical analysis of *Papaver* alkaloids by immunoassay. Most of these made use of commercial antibodies [143, 305, 338]. E.g., the appearance of morphinan alkaloids in correlation with a structural study of lacticifer cell appearance in germinating *P. bracteatum* seedlings [305] was based on an immunological assay using a commercial anti-morphine antibody. Hsu et al. [151] have used a RIA, based on antisera elicited against morphine-3-hemisuccinyl-BSA, to quantitate morphine (*33*) in extracts of *P. somniferum* capsules. However, the usefulness of the assay was reduced due the cross-reactions of codeine (*34*) and thebaine.

This problem was recently overcome by Zenk's group who have synthesized the N-2-carboxyethyl derivatives of (S)- and (R)-nor-reticuline, northebaine, norcodeine and normorphine as well as the salutardine-7-(O-carboxymethyl)oxime, coupled them to BSA, and for each compound, elicited exceptionally selective antisera [388,389]. The sera were used in an RIA to determine opium alkaloids in herbarium leaf samples of *Papaver* species, in *P. somniferum* individuals in breeding experiments and the formation of opium alkaloids during germination [389]. Moreover, the antibodies now available are powerful probes in studies on the biosynthetic pathway to these benzylisoquinoline alkaloids.

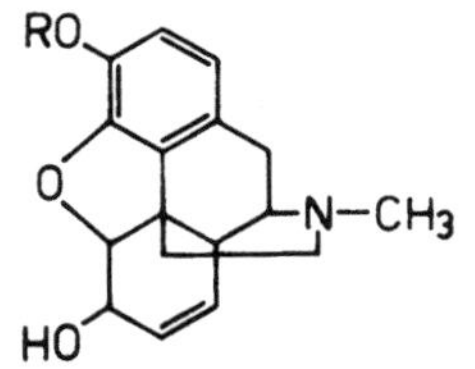

(33) R = H

(34) R = CH_3

Other Alkaloids

Theophylline (1,3-dimethylxanthine) (*35*) is a natural alkaloid which is used extensively for the prevention and treatment of asthma and for the treatment of juvenile bradycardia. Since the compound has a narrow therapeutic range, immunological monitoring during therapy was of interest and a whole range of immunoassays is now available including the EMIT® technique [120], RIA [54,246], fluorescence polarisation immunoassay [144,206], and solid phase reagent strip immunoassay [206a,355], and a highly sensitive immunoliposome assay [111a].

Colchicine (*36*) antibodies were produced via N-desacetylthio-colchicine with free, carbodiimide-activated carboxyl groups of BSA and used to detect ng amounts of the compound, by RIA [24].

The *Cinchona* alkaloids, quinine (*37*) and quinidine (*38*) rank among the 15 most prescribed drugs of plant origin. The anti-malaria drug, quinine (*37*), is also used as a bittering agent by the soft-drink industry and its stereoisomer, quinidine (*38*), is an effective anti-arrhythmic drug. The compounds are obtained by extraction of the bark of mature *Cinchona* trees. Their production in plant tissue culture would be highly desirable in order to secure a constant supply of the drugs and to spare natural resources. Robins et al. have produced, via 9-hemisuccinylquinine, antisera selectively recognizing quinine (*37*) and closely related 8S,9R alkaloids. No cross-reactions were observed with the 8R, 9S steric group (quinidine (*38*) and related compounds) [300]. This RIA detects ng amounts of quinine (*37*) and has been applied to the estimation of alkaloid content of clonal shoot cultures of *Cinchona ledgeriana*. More recently, an ELISA for pg amounts of quinine (*37*) and RIA and ELISA techniques for quinidine (*38*) were reported from the same group [236]. Quinidine (*38*) immunoassays are also commercially available [260,287] but have not been applied to plant analysis. A fluorescence immunoassay for quinine (*37*) was recently reported [316a]. The quinolizidine alkaloids, lupanine (*38a*), 13-hydroxylupanine and related compounds, have been determined in plant tissue cultures by an RIA with a detection limit of 0.05 ng of alkaloid per assay. Only an abstract of the work is available [393a)].

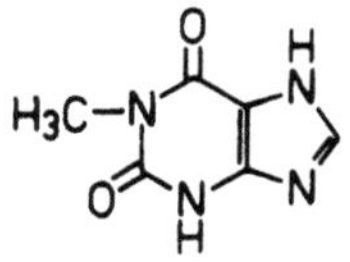

(35)

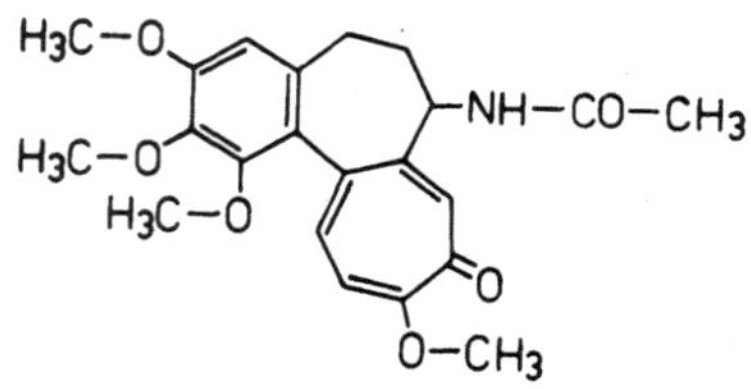

(36)

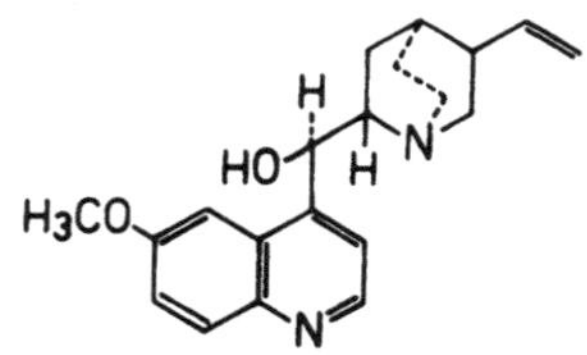

(37)

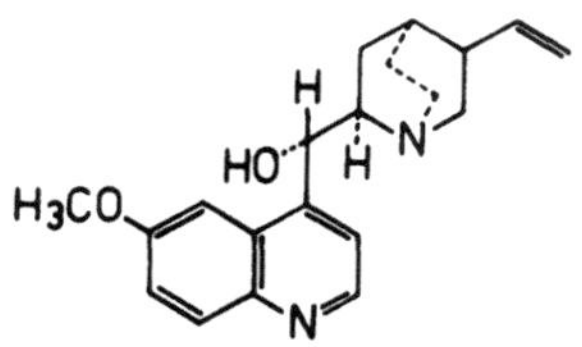

(38)

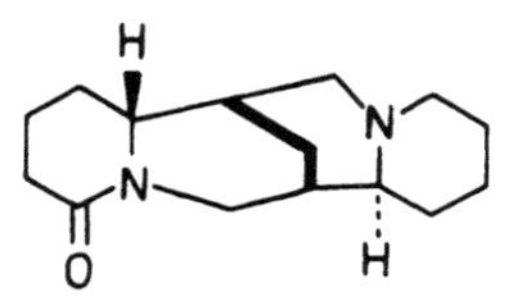

(38a)

4.2.3 Miscellaneous Secondary Metabolites

In addition to the groups of plant natural products discussed above, immunochemical approaches were used to detect or quantitate members of the following classes of compounds: biogenic amines, flavonoids and isoflavonoids, anthrone glucosides, and glucosinolates.

Anthrone Glucosides

The dianthrone glucosides, sennoside A and B (*39*) possess strong laxative properties and are extracted from various species of *Cassia* and *Rheum.* In a pioneering study for this class of compounds, Atzorn et al., [11)] have investigated the immunological properties of the sennosides A and B (*39*). These differ only in the stereochemistry of the $C_{10}—C_{10'}$ link (Sennoside A: (+), Sennoside A_1 : (—), Sennoside B (*39*) (meso)). Since it was important to distinguish the glucosides from the aglycones (sennidines), the 3′- and 3-carboxyl group were used to link sennoside B (*39*) to BSA. Sera were obtained from rabbits and used in a RIA in conjunction with ^{3}H-sennoside B synthesized from (^{3}H)-8-glucosidorheinanthrone. The assay proved very selective for (a) the glycosidic compounds and (b) for the stereochemistry of the $C_{10}—C_{10'}$ linkage, but rhein-8-glucoside and 8-glucosidorheinanthrone, monomeric precursors of sennosides, were also detected. The assay was used to study the distribution of sennoside B (*39*) in various parts of *Cassia angustifolia* including fruit samples. A kinetic study of the levels of sennoside B (*39*) in early stages of seedling development was also conducted. In addition, it was shown that the dianthrone, sennoside B (*39*), is apparently formed under the oxidative fermentation conditions of the leaf drying process from monomers. The substance is nearly absent in rapidly extracted fresh leaf (*C. angustifolia*).

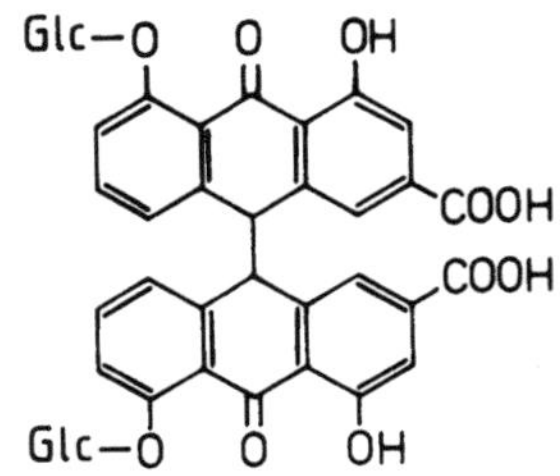

(39)

Flavonoids and Isoflavonoids

Naringin (*40*) (naringenin-7-O-α-rhamnosyl-(1–2)-β-D-glucopyranoside) is an intensely bitter flavanone glycoside of grapefruit. Bitterness is associated with the neohesperidosyl moiety, since the isomeric rutinosyl moiety (α-O-rhamnopyranosyl-(1–6)-β-D-glucopyranoside), also occurring in grapefruit, is non-bitter. Excessive bitterness due to the presence of naringin (*40*), but also limonin (*13*) (see Chapt. 3.2.1.) adversely affects the quality of the fresh fruit and the processed juice.

A systematic study on the immunoassay determination of naringin (*40*) was carried out in our and particularly in Mansell's laboratory [164,165,166,167]. Immunogens of naringin (*40*) were synthesized through the naringin-4-O-carboxymethyloxime and through azo links between diazotized, *p*-aminohippuric acid substituted BSA, and the naringin-B ring. Antisera raised against naringin-4-O-carboxymethyloxime-BSA conjugates, in conjunction with (^{3}H)naringinol of high specific activity gave best results in terms of assay sensitivity (detection limit: 0.3 pmol/assay [167]). As could be expected from the immunogen structure, the neohesperidosyl moiety proved an essential component for immunological reactivity. Thus, the assay is useful to quantitate, at low levels, a range of neohesperidosides such as neohesperidin, poncirin, rhoifolin or fortunellin after TLC separation. It does not react with the aglycone, naringenin, or with isomeric glycosides such as rutinosides. This is an excellent example of group-selectivity in immunological assays. These assays have been applied to the large scale screening of naringin (*40*) presence in leaf samples of *Citrus*[167], the determination of naringin (*40*) were synthesized through the naringin-4-O-carboxymethyloxime and content and distribution in *Citrus paradisi* at all stages of vegetative and generative development and in all parts of the plant [165] including callus cultures [16 a)]. These studies beautifully exemplify how immunoassay can help to gain a very thorough understanding of natural product formation and accumulation during the life cycle of a plant. This study was recently extended to include a radioimmunoassay for hesperidin (*40a*) [166]. The 4-0-carboxymethyloxime of hesperidin, coupled to BSA, served as immunogen, and borotridide-reduced hesperidin-4-alcohol as radiotracer. The assay was applied to the analysis of hesperidin distribution in leaves, fruit, seeds and seedlings of *Citrus sinensis*.

Recently, antisera were developed for the partially methylated flavonol glucoside, 5,2′,5′-trihydroxy-3,7,4′-trimethoxyflavone-2′-O-glucoside (*41*) [183] and used for the immunofluorescence localization of the glycoside (*41*) in *Chrysosplenium americanum*[29].

As in the preceding cases [164,167,183] diazotized *p*-aminohippurate-substituted BSA was used by Moesta et al. [235] to synthesize an immunogenic conjugate of the isoflavonoid, pterocarpane phytoalexine, glyceollin I (*42*) which allowed the quantitation by RIA, of pmol amounts of the phytoalexin in *Phytophthora megasperma* infected soybean tissue [114,235].

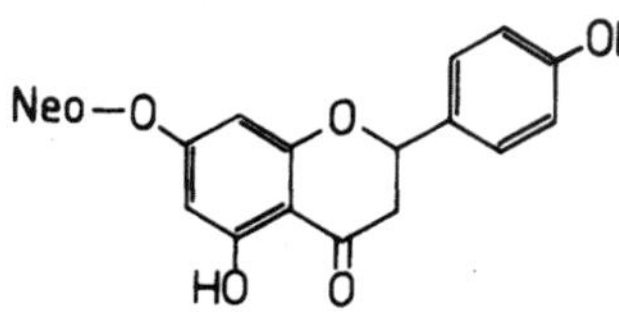

(40)

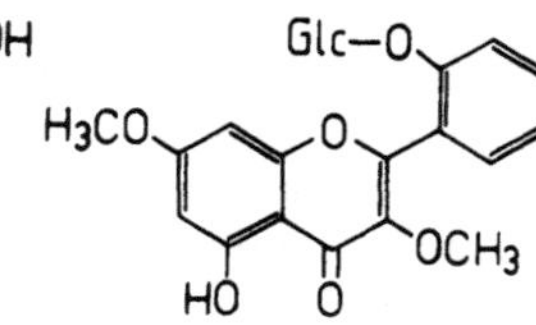

(41)

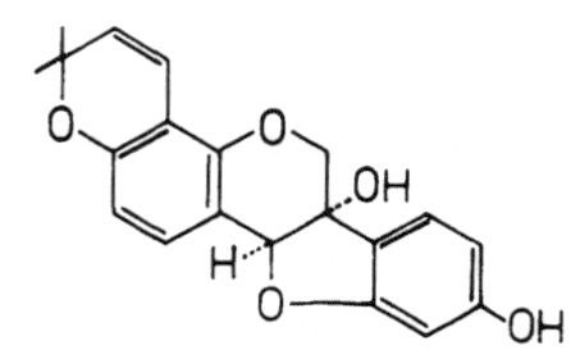

(42)

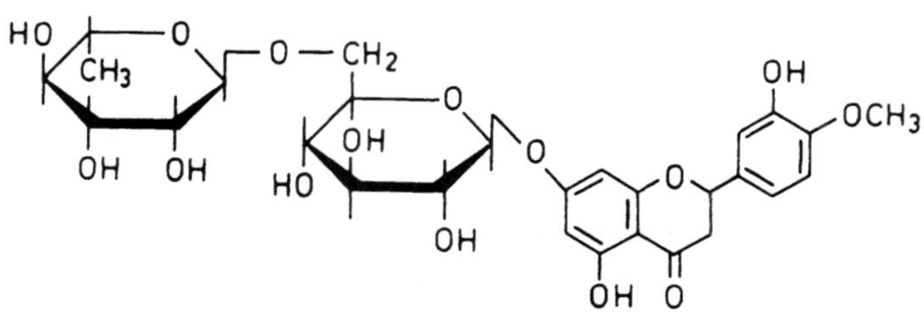

(40a)

Biogenic Amines

In this section, we briefly discuss some selected references on immunoassays for biogenic amines of the catecholamine or polyamine type. Some of these compounds are rather simple in structure and the main purpose for including them in this review is to exemplify the fact that specific antisera can be raised against very small antigens. Tyramine (*43*) is chosen here as a first example of an antigen of the catecholamine group. This compound is a naturally occurring sympathomimetic amine in mammals but is also present in plants. Faraj et al. [92] used diazotized *p*-aminohippuric acid-substituted BSA to link tyramine (*43*), through an azo link in the 4-position. This left the aminoethyl side chain exposed. Their RIA detects as little 1 ng tyramine (*43*) and is rather specific (e.g. there is a sharp discrimination between tyramine (*43*) and phenylethylamine, dopamine or L-tyrosine). In a different approach [119] the catecholamines tyramine (*43*) and synephrine (*44*) were conjugated to BSA by formaldehyde condensation, and β-3,4-dimethoxyphenyl-ethylamine was converted to the N-succinyl derivative which in turn was coupled to poly L-lysine [296]. In both cases, high affinity antibodies (sera) were obtained which proved useful for RIA. But as expected, side-chain variants showed considerable interference. (e.g. 3,4-dimethoxybenzylamine in the assay of 3,4-dimethoxyphenylethylamine (*45*) [296]. Interestingly, the formaldehyde condensation technique [119] which should couple through the amino group, both in the case of tyramine (*43*) and in the case of synephrine (*44*) resulted in antibodies which bound 3-methoxy analogues with much higher affinity than the original antigen. Most recently, highly selective antisera against the cathecholamines, L-epinephrine ([399 c], and refs. cited therein), both structurally related to the *Ephedra* alkaloid, ephedrine, were raised and used in RIA analysis of these compounds. Of importance to those interested in plant phenolics or catecholamines will be a recent report, from the same group, of the preparation of monoclonal antibodies against vanilmandelic acid and homovanillic acid [399 a] which were raised against immunogens prepared by condensation of vanilmandelic acid or homovanillic acid with primary amino groups of HSA. Condensation proceeded at the 5-position of the aromatic ring, through a methylen bridge from formaldehyde (Mannich reaction) [399 a].

Recently, monoclonal anti-histamine antibodies were raised using histamine, coupled through 1,4-benzoquinone to OVA. The histamyl-benzoquinone hapten exhibited a 2.5×10^5 fold greater affinity than free histamine [44 a] and consequently, free histamine had to be converted to the benzoquinone derivative prior to analysis. This is another example which demonstrates the drastic influence the bridge-structure may have on the binding of haptens vs. underivatized LMC to their antibodies.

Polyamines, such as spermine (*46*) and spermidine (*47*), occur widespread in animals and plants. In plants, they have growth promoting properties, probably through their stabilizing interaction with DNA and membranes [318]. Specific spermidine (*47*) antibodies were produced by immunizing rabbits against a spermidine thyroglobulin conjugate with high coupling ratio (200 hapten moieties per protein) [17]. The antiserum was very selective in that spermidine (*46*) cross reacted to only 0.3%, cadaverine to 0.4% and the precursor, putrescine, to only 2.1% (spermidine: 100%). By using (^{3}H)spermidine as tracer, it was possible to quantitate the amine in the 1–10 ng range. This RIA was not applied to plant analysis.

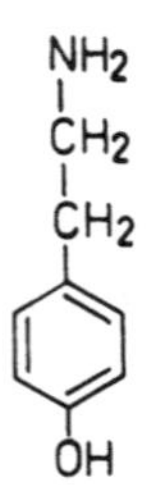

(43)

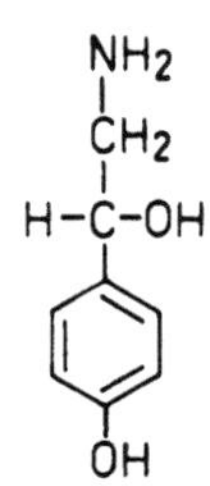

(44)

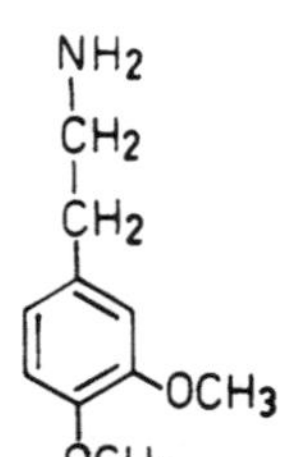

(45)

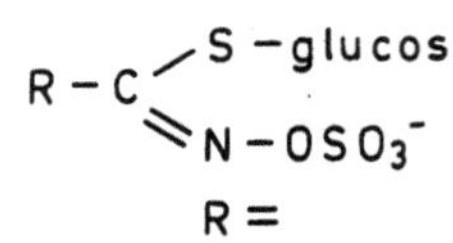

R =	
(47a)	$CH_2{=}CHCH_2-$
(47b)	$CH_2{=}CHCH_2CH_2-$
(47c)	$CH_3CHCH(OH)CH_2-$

$NH_2{-}(CH_2)_3{-}NH{-}(CH_2)_4{-}NH_2$

(47)

$NH_2{-}(CH_2)_3{-}NH{-}(CH_2)_4{-}NH{-}(CH_2)_3{-}NH_2$

(46)

Glucosinolates

The presence of glucosinolates presents an obstacle to the use of oilseed such as rapeseed and Indian mustard as stock feed and considerable effort is being spent to reduce the glucosinolate content of these plants.

In a pioneering effort on this class of compounds, Hassan et al. [134a)] have developed an ELISA for the alkenyl glucosinolates, sinigrin (*47a*), gluconapin (*47b*) and progoitrin (*47c*). Immunogens were synthesized by coupling sinigrin, though the glucosyl-moiety, to BSA or OVA. Sinigrin was first converted to the 6'-mono-hemisuccinate which in turn was converted to the N-hydroxysuccinimide ester. The active ester of sinigrin was reacted with protein-NH_2 [134a)]. Antisera raised in rabbits against the BSA-conjugate were of high titer and allowed to detect as little as 0.1 pmol of sinigrin by ELISA. Cross reaction with the related glucosinolates, gluconapin (*47b*) and progoitrin (*47c*) was high. In Indian mustard, *Brassica juncea*, progoitrin is a minor glucosinolate, whereas sinigrin and gluconapin are the dominant compounds, occurring in varying relative amounts in different varieties. The identical cross-reactivity of the two glucosinolates allows to quantitate their total amount directly in, e.g., seed meal extracts and extracts from vegetative tissues. Validation experiments have shown that ELISA and a HPLC-based technique for glucosinolate determination yield identical results within the limits of experimental error, however, the coefficients of variation were lower (3% as compared to 9.5%) for the ELISA technique than for the HPLC-assay [134a)].

5 Immunoassays for Endogenous and Synthetic Plant Growth Regulators

5.1 Scope

The quantitation of plant growth regulators still represents one of the biggest challenges of phytochemical analysis. Levels of growth regulators hardly ever exceed 1 μM, but may be as low as 10^{-9} to 10^{-10} M in many plant tissues. A wide spectrum of techniques has been developed to determine the levels of growth regulating compounds. These include bioassays and physico-chemical analysis and, more recently, immunological techniques based on polyclonal as well as monoclonal antibodies. Bioassays and physico-chemical analysis of growth regulators as well as the general problems associated with their estimation were discussed recently in several reviews [27,148,293)] and growth regulator immunoassays were reviewed in [361,374,375)]. The technical aspects of the use of growth regulator immunoassays are discussed especially in [375)]. From the experience with RIA as well as ELISA techniques for plant growth regulators, the following generalizations can be made:

a) Immunoassays are, under almost any circumstances, more sensitive than or equally sensitive to alternative techniques. The potential inherent in immunoassay is exemplified by the recent development of a microscale, enzyme amplified immunoassay for abscisic acid (ABA) (*48*) which allows the quantitation of as little as 200 attomoles of ABA (*48*). This represents the hormone level present in 10 pairs of guard cells of bean epidermal tissue [133]. The sensitivity of this assay opens completely new dimensions for the analysis of growth regulators in plants.

b) Immunoassay almost always allows a reduction in the number of clean-up steps required prior to the quantitation step, although there are exceptions [51 a)]. In some cases, purification was not necessary at all, whereas, for other tissues, quite extensive clean-up was required despite the use of rather specific antibodies (e.g., [51 a)]). At present, general rules for pre-analysis processing of samples cannot be given. Compared to alternative techniques, clean-up in most cases is considerably less for immunoassay. This results in savings of experimental time, reduces work-up losses and allows one to analyse very small samples for several growth regulators at the same time.

Immunoassays for growth regulators were designed primarily to quantitate those compounds in tissue extracts and to study the dynamics of hormone levels in a variety of physiological and pathological situations. Furthermore antibodies were used to purify growth regulators from tissues, or from synthetic materials, by immuno-affinity chromatography and also to localize them within tissues and cells, by immunocytochemical techniques. The earliest reported use of antibodies to study growth regulators dates back to 1969 [102], but these pioneering studies were, at that time, largely ignored. Interest grew again after Pengelly et al. [266] and Weiler [380] introduced RIA to this field and since then an ever increasing number of reports document the growing acceptance of these techniques by the scientific community. In the following chapters, a detailed discussion of the available techniques is presented. Table 2 gives a summary account of the immunogen structures used to elicit immune responses against plant growth regulators while Table 3 surveys radio- and enzyme-tracers and Table 4 presents the assay formats and general parameters of a selection of representative immunoassays for the quantitative determination of plant growth regulators.

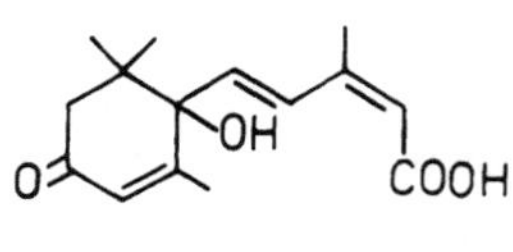

(48)

5.2 Auxins

The first antibodies directed against indole-3-acetic acid (IAA) (*49*) were reported by Fuchs and Fuchs in 1969 [102] and Fuchs et al. in 1971 [103]. These authors produced an immunogen by coupling IAA (*49*) to hemocyanin through the carboxyl function. The antisera were used to measure IAA (*49*), at a detection limit of ca. 60–300 pmol ml^{-1} in complement fixation and T4 phage inhibition assays [103]. Specificity data and applications to plant material were not reported. The RIA technique was introduced to IAA analysis by Pengelly and Meins [266]. This assay

Table 2. Structures of immunogenic plant growth regulator-protein conjugates. n.g., not given

Hormone	Coupling site	Coupling procedure	Carrier	Refs.
IAA (*49*)	N-	Mannich reaction	BSA	262 a, 266
	carboxyl	Mixed anhydride	BSA	383
	carboxyl-	Via ε-aminocaproate	KLH, BSA	103
(±)-ABA (*48*)	carboxyl-		KLH	105
	carboxyl-	EDC	HSA	380
	keto-	Complex spacer via ABA-4′-tyrosylhydrazone	BSA	381
	keto-	Mixed anhydride via methyl-ABA-(O-carboxymethyl)oxime	THY	304 a
(+)-ABA	carboxyl-	EDC	BSA, HSA	202, 381
	carboxyl-	Mixed anhydride	BSA	379
	keto-	Complex spacer via diazotized ABA-4′-(p-aminobenzoyl)-hydrazone	KLH	286
	keto-	Complex spacer via ABA-4′-Nα-acetyl-lysine hydrazide	BSA	286
Cytokinins	ribosyl-	Via dialdehyde after per-iodate oxidation	BSA, KLH	53, 88, 130, 147 356a, 382, 368, 401
trans-Z (*53*)	N-9-	Michael addition through 9-carboxyethylzeatin	n.g.	26 a
GA1 (*57*)	carboxyl-	Mixed anhydride	BSA	12
	hydroxyl(C3)-	Mixed anhydride via 3-hemisuccinate	KLH	180 a
GA3 (*58*)	carboxyl-	Via ε-aminocaproate	BSA, KLH	103
	carboxyl-	Mixed anhydride	BSA	371
GA4 (*59*)	C-17-	Via 16,17-dihydro-17-yl-1′-(1′-thia) butane-4′-thiol through succinic anhydride	KLH, BSA	180 a
	carboxyl-	Mixed anhydride	BSA	14
	C-19/C-20-	via GA13-19,20-imide-β-alanine-7-methyl ester	BSA	79
GA5 (*60*)	carboxyl-	Via GA5-7-β-alanine	BSA	399
GA9 (*62*)	carboxyl-	Mixed anhydride	BSA	12
	C-17-	Via 16,17-dihydro-17-yl-1′-(1′-thia)butane-4′-thiol through succinic anhydride	KLH, BSA	180 a
GA20 (*63*)	carboxyl-	Via GA20-7-β-alanine	BSA	399
GA24 (*65*)	C-17-	Via 16,17-dihydro-17-chloroformate	BSA	181
	C-17-	Complex spacer via 16,17-dihydro-17-cyanoacetate	BSA	181
(±)-Jasmonic acid (*67*)	carboxyl-	EDC	BSA	179
2,4-D (*50*)	C-5-	Via diazotized 5-amino-2,4-D	BSA	297

Table 3. Structures of tracer plant growth regulators for use in RIA and ELISA. The following abbreviations were used; T, tritium; I, iodine-125; AP, alkaline phosphatase; n.g., not given.*, in TBq/mol for radiotracers, AP used typically at a specific activity (Spec. Act.) of over 2000 U/mg protein)

Derivative	Label	Position of label	Typical Spec. Act.*	Refs.
A. Radioactive phytohormones and derivatives:				
IAA (*49*)	T	C-5	1000	266
IAA-tyrosine methyl ester	I	Benzyl C-3, C-5	15000	383
(±)-ABA (*48*)	T	Mainly side-chain	830	356
(±)-ABA-gly-tyr	I	Benzyl C-3, C-5	n.g.	194
Cytokinin-riboside dialcohols	T	Oxidized ribose	400—800	382
trans-ZR-tyramine	I	Benzyl C-3, C-5	19000	78
(±)-DHZ (*55*)	T	Side chain	1600	147
GA1 (*57*), GA4 (*59*)	T	C-1, C-2	1400—1600	180a
GA5-, GA20-methyl ester	T	Methyl group at C-7	2600	399
GA9 (*62*)	T	C-2, C-3	1800	180a
GA3-17-nor-16-ol	T	C-16	220	12
GA9-17-nor-16-ol	T	C-16	180	12
GA3-7-[N-(p-hydroxybenzyl)-putrescine]amide	I	Benzyl C-3, C-5	85000	371
(±)-Jasmonic acid (*67*)	T	Uniform	7.4	179
B. Enzyme-labeled phytohormones and hormone derivatives:				
IAA (*49*)	AP	N-	n.g.	262a
	AP	Carboxyl-	n.g.	363
(±)-ABA (*48*)	AP	Carboxyl-	n.g.	63
	AP	Keto-	n.g.	375
(±)-ABA (*48*)	AP	Carboxyl-	n.g.	379
Cytokinin ribosides	AP	Ribosyl-	n.g.	77, 78
GA3 (*58*), GA4 (*59*)	AP	Carboxyl-	n.g.	14

represented a considerable improvement over the earlier techniques in that it was more sensitive (level of detection ca. 1 pmol IAA (*49*) per assay) and made use of a highly specific antibody raised against IAA-BSA conjugates condensed by the Mannich-reaction. Since then, an array of immunological assays for IAA quantitation has been reported employing either N-coupled [262, 262a, 266, 329)] or side chain coupled [218, 220, 262, 306, 347, 350a, 360, 363, 383)] IAA as immunogens. While N-coupling leaves the characteristic side chain of IAA (*49*) maximally exposed and results in a highly polar haptenic structure, carboxyl-coupling masks the side chain and results in a less polar haptenic structure. Several workers observed higher and more consistent titers for the latter type of immunogen while N-coupled IAA apparently created only a weak (or unstable) immunogen, resulting in a low and highly variable immune response [266, 383)]. On the other hand, only the latter type of sera reacted with deprotonated IAA (IAA^-) which predominates at normal assay pH while antibodies elicited against side chain coupled IAA require chemical modification prior to analysis (removal of the charge at C1, mimicking of the hapten-typic "bridge" structure). This chemical modification is conveniently done by methylation of IAA

Table 4. A selection of representative plant growth regulator immunoassay formats. The following abbreviations are used: R, RIA; E, ELISA; F, FIA; EA, enzyme amplified immunoassay; SP, solid phase assay; LP, liquid phase assay; T, tritium; AP, alkaline phosphatase; GD, glucose-6-phosphate dehydrogenase; M, monoclonal antibody; P, polyclonal antiserum; W, within assay reproducibility (as % coefficient of variation), B, between assay reproducibility, (as % coefficient of variation); (me), used as the methyl ester

Standard	Assay format	Measuring range [pmol/assay]	Reproducibility for standards	Refs.
IAA (me)	E,M,SP,AP	1—30	W 2	230
cis(+)ABA	R,M,LP,T	0.02—2	W 2—5	229
	EA,M,SP,AP	0.0002—0.01	W 5—10	133
Xanthoxin	R,P,LP,T	1—200	W 5—6, B 6	95a
GA1 (me)	R,P,LP,T	0.07—5	W 4—5	12
GA4 (me)	R,P,LP,T	0.2—10	W 1—5	79
GA5 (me)	R,P,LP,T	0.03—10	W 5	399
GA9 (me)	R,P,LP,T	0.1—5	W 4—5	12
GA20 (me)	R,P,LP,T	0.03—10	W 5	399
GA24 (me)	R,P,LP,T	0.05—10	n.g.	181
IPA	E,M,SP,AP	0.1—5	n.g.	78
DHZR	E,M,SP,AP	0.02—10	W 1	77
trans-ZR	E,M,SP,AP	0.02—10	W 2	77
2,4-D	E,P,LP,GD	—	n.g.	94a
Maleic hydrazide	E,M,SP	0.17—17 ppm	—	134b

or samples [363,383]. In several cases, the IAA ethyl ester showed a dramatic increase in affinity compared to IAA methyl ester, however, the ethylation reaction requires even more rigid safety precautions than methylation and is thus impractical. Antisera raised against N-coupled IAA are considerably more selective than antisera raised against side chain coupled IAA, especially for variations in the side chain [266]. This is important because of the occurrence of high levels of IAA esters and peptidyl conjugates in several plant tissues. On the other hand, the need for sample methylation has the advantage of chemically stabilizing IAA (*49*) which, in the free form, is extremely labile in the presence of light and oxygen. Consequently, workup losses are reduced and the reproducibility of the analysis is improved.

More recent advances in the immunoassay of IAA (*49*) include the introduction of the EIA technique based on immobilized antibody [230,363] or immobilized antigen [220], the use of avidin-biotin coupling of the tracer reaction to the immune reaction [220] and the generation of high affinity monoclonal antibodies [230]. It is now possible to quantitate around 3–5 pg of IAA (*49*) immunologically and this is quite close to the sensitivity of the fluorescent assay for IAA (*49*) [57]. Sundberg et al. [329] have successfully employed immobilized antibodies for the immunoaffinity purification of IAA (*49*) from plant extracts (e.g. *Pinus sylvestris*) prior to HPLC analysis. IAA was efficiently enriched allowing for direct quantitation from the HPLC trace of the affinity column eluted material [329].

Immunological analysis of IAA (*49*) was used in physiological studies, to analyze hormone distribution during gravitropic response [233], hormone levels in phloem exudates [373], fluctuations of IAA (*49*) levels in epidermal tissue of *Commelina* during the day/night cycle [289], IAA (*49*) levels in various plant-tissues such as pine needles [89a, b, 329], spruce needles [350a], the vascular cambium of *Abies balsamina* [329a], maize caryopses and seedlings [264, 265, 363], *Nicotiana* pith parenchyma and teratoma tissue [266], crown gall tumors [4, 368, 378, 398], *Pisum* roots [263], sunflower roots in relation to boron supply [141], tomato roots [221], and *Drosera* leaf blades after capture of a prey [23], coleoptiles [363] and phytopathogenic bacteria [368] as well as mycorrhizal fungi [218]. An interesting study on the hormone status of the avocado plant made use of immunoassays for several growth regulators including IAA (*49*) [59, 61, 394]. This study exemplifies the advantageous use of immunological techniques to establish the hormone status of important crop plants thus permitting a trend analysis for correlating changes in the hormone status with certain physiological events. Such knowledge is an important prerequisite in order to manipulate plant growth and development, but may also provide predictive measures of certain late processes of development, e.g. crop yield or fruit set during crop growth. Similar studies were recently carried out to follow changes in the levels of several endogenous plant growth regulators in soybean seedlings following the application of the growth retardants, tetcyclaeis and LAB 150978 [118a], in tissue cultures of *Pennisetum purpureum* [289a] as a function of culture-stage and in florets of single wheat spikelets during pre-anthesis development [192a].

Using immunoassays based on poly- [363] and later monoclonal antibodies [230] to determine enzymatic activity, Schröder and colleagues were able to elucidate the function of one of the *Agrobacterium tumefaciens* onc genes as an amidohydrolase which converts indole-3-acetamide to IAA (*49*) [174, 311]. This too (cf. Chapt. 3.2.2.1) exemplifies the power of immunological assays for enzymatic studies in cases when enzyme assays are not available or the precise activity is not known. With the interest in chemical growth regulation increasing, monitoring of the endogenous growth regulator levels during treatment will become more and more important. Immunoassays may prove particularly helpful in such studies and this potential is rapidly being recognized [117]. In most systems tested, immunological techniques for IAA (*49*) have proven both accurate and precise, [264, 265, 306]. However, due to the presence of sometimes excessive amounts of structurally related indoles and/or IAA conjugates in some tissues, a pre-cleaning of samples prior to analysis will be required, including HPLC steps [51a]. An attractive alternative, immunoaffinity purification of IAA followed by physicochemical quantitation, was recently shown to be applicable to plant analysis [329]. If MAB recognizing the deprotonated, free hormone became available, this would further increase the attractiveness of the immunological assay for IAA (*49*). Although a range of other auxins occur in plants, immunoassays for their detection have not been reported to date.

Rinder and Fleeker have described the RIA of the synthetic auxins, 2,4-dichlorophenoxyacetic acid (*50*, 2,4-D) and 2,4,5-trichlorophenoxyacetic acid (2,4,5-T) which are used as herbicides. They used 5-amino-2,4-D as an intermediate of immunogen synthesis and were able to elicit sera recognizing 2,4-D (*50*) and (better) 2,4,5-T. Other related compounds were recognised to a much lesser extent (e.g. 2,4-dichlorophenol cross reacted 0.2%, 2-(2,4,5-trichlorophenoxy)propionic acid 7.8% and 2-methyl-4-chlorophenoxy acetic acid 2.4% relative to 2,4-D). The assay was used as an initial screen to detect 2,4-D (*50*) and/or 2,4,5-T in samples of surface water after the pesticide was concentrated on silica-C_{18} cartridges [297]. A further simplification of 2,4-D (*50*) analysis might come from the introduction of a homogenous enzyme immunoassay (EMIT®) for this compound which detects as little as 2.5 ng of 2,4-D (*50*) [94 a)]. The assay has not yet been applied to residue analysis.

The synthetic growth regulator, maleic hydrazide (1,2-dihydro-3,6-pyridazinedione) (*50a*) is discussed in this chapter because of its many auxin-like effects. The compound is used as a growth retardant for grass and ripening retardant on apple trees. It strongly inhibits the outgrowth of axillary buds and this action is exploited to prevent sprouting of onions and potatoes during storage as well as to prevent sucker formation in detopped tobacco. Harrison et al. [139 b)] have coupled acetic acid derivatives of maleic hydrazide (*50a*) to carrier proteins, immunized mice and derived hybridomas secreting anti-maleic hydrazide MAB, albeit of rather low affinity (1.9×10^5 M^{-1}). One MAB allowed to detect as little as 0.5 ppm of the growth regulator added to filtered potato juice and produced a linear response from 0.17 to 17 ppm of standard maleic hydrazide (*50a*). The MAB cross-reacted with three of the four maleic hydrazide derivatives tested, including maleic acid-β-D-glucoside, a metabolite found in tobacco.

Recently, a RIA for 5-hydroxy-IAA was described using antisera raised against 5-hydroxytryptamine succinamide-protein conjugates [225 a)].

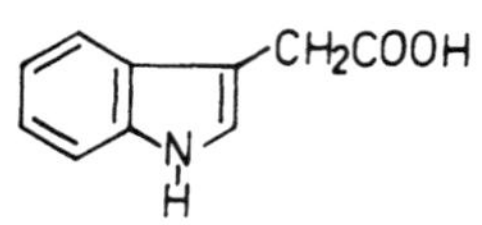

(49)

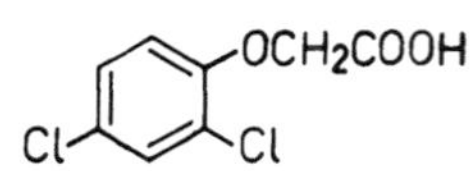

(50)

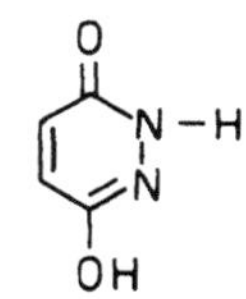

(50a)

5.3 Cytokinins

The immunoassay of cytokinins was recently reviewed in detail [89)], therefore only the most important aspects will be summarized here. For the major groups of endogenous free cytokinins, antisera and monoclonal antibodies of high quality are now available. The technique used for immunogen synthesis in most studies was basically the same: starting from periodate oxidized cytokinin-9-riboside, the resulting di-aldehyde was reacted with protein under basic conditions (pH >9) and the Schiff's bases formed were stabilized by reduction with sodiumborohyride. This reaction was first described by Erlanger and Beiser [84)] to generate anti-nucleoside antibodies, then later adopted for the synthesis of digoxin-immunogens as well as for other glycosidic compounds (see Chapt. 3.2.1). In a different approach [26 a)], the 9-carboxyethyl derivative of tZ was synthesized and this coupled to protein. The MAB obtained from animals immunized against this conjugate were compared to those obtained using ribosyl-coupled tZR immunogens. tZR, in all cases, effectively bound to the MAB, allowing their use in ELISA based on enzyme conjugates of tZR.

Cytokinin-protein conjugates are highly immunogenic and elicit, in rabbits and mice, a uniform and strong immune response. It is particularly impressive to realize that several laboratories using different species of animals and different assay configurations, but the same synthetic procedure for the synthesis of the immunogen, have all reported very similar antibody characteristics (see comparison in [15)]). This exemplifies the dominant and also consistent influence of immunogen structure on the resulting immunological response. RIA [15,15a,62,67,88,147,175,215,217,234,261,342,352,368,382,396], ELISA [16,26a,36,77,78,105a,128,146,220,322a,354] and FIA [340a] techniques are available for the quantitation of cytokinins of the isopentenyladenine- (*51*, *52*) [15a,16,36,62,67,78,88,105a,128,146,175,215,234,322a,340a,354,357,368,369], cis-zeatin- [261], transzeatin- (*53*, *54*) [15,16,36,105a,128,146,175,215,220,294,340,340a,341,342,354,357,382] and dihydrozeatin (*55*, *56*) group [15a,77,105a,147,309]. ELISA permits the detection of as little as 0.1 pmol of a given cytokinin. The availabilities of monoclonal antibodies directed against key members of all major groups of cytokinins [26a,77,78,216,340,340a,341,356a,396] was a prerequisite for widespread use of such assays. Due to the peculiar immunogen structure employed, both the free cytokinin bases, and also the corresponding 9-ribosides and 9-ribosyl-5′-phosphates are highly immunoreactive. This has two consequences: i) if no separation of these metabolites is carried out, then only a gross estimate of cytokinin immunoreactivity (e.g. trans-zeatin type reactivity etc.) can be made, ii) if an absolute quantification is desired, then the free bases, ribosides and ribotides have to be separated prior to analysis, but the members of one cytokinin group can be analyzed with a single immunoassay. These include the cytokinin ribosyl-5′-phosphates which with other assays can only be analyzed after conversion to the ribosides.

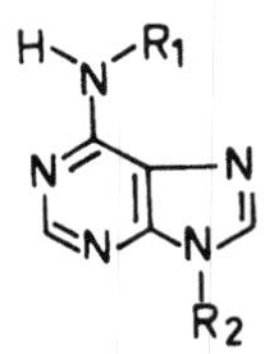

	R_1	R_2
(51)	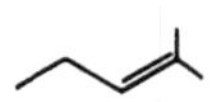	H
(52)	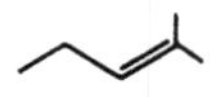	Rib
(53)	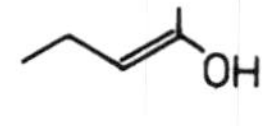	H
(54)	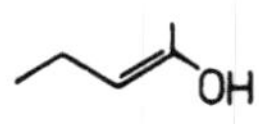	Rib
(55)	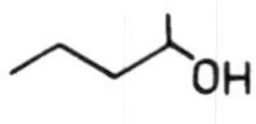	H
(56)	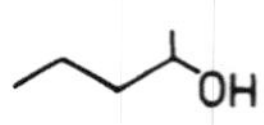	Rib

Applications: Cytokinin immunoassays have already been applied to a wide range of plant and microbial systems. These applications include the following examples: determination of cytokinins during crown gall formation and in crown gall cell cultures [4,245,294,368,378,398], cytokinin production by phytopathogenic bacteria, such as *Agrobacterium tumefaciens* [294,368], assays for dimethylallyl-transferase activity in studies on the molecular biology of the crown gall disease [138], hormone levels during adventitious root formation [221], during Avocado growth and development [59,61,394], in the phloem exudate of several tree species [373], in maize seed [142 a)] and during maize germination [142] and in maize seedings [130], anise cell cultures [86,88], tobacco cell cultures [346 c)], *Citrus* blossom development [16], and to study the *in vitro* metabolism of dihydrozeatin by soybean callus [147]. Such assays were also used to analyze cytokinin levels in seaweeds [146], Douglas fir [244] and wheat ear [75], wild type and cytokinin overproducing mutants of the moss, *Physcomitrella patens* [357], root pressure exudates of *Urtica dioica* [105 a)], xylem exudates of *Phytophtora cinnamomi* infected *Eucalyptus marginata* [36] and dormant potato tubers [342]. Of considerable interest is the use of immobilized cytokinin antibodies for the single-step enrichment of analytical-quality cytokinin fractions from complex plant or microbial samples. Immunoaffinity purification to achieve this goal was used in several studies (see Chapt. 4.6) and proved exceedingly powerful, e.g. leading to sufficiently clean samples from crown gall extracts or culture filtrates of *A. tumefaciens* to allow physico-chemical quantification after a single HPLC step [217]. For culture filtrates of *Physcomitrella patens*, a comparison between RIA and GS-MS showed that RIA yielded accurate results when the unprocessed culture filtrates were analyzed [357]. In this case, only the cytokinin free bases are secreted, thus, interference from ribosides and/or ribotides is excluded.

Synthetic cytokinins: Only very few studies have been carried out with synthetic cytokinins. Constantinidou et al. [53] reported the production of an antiserum against N^6-benzyl-adenosine, a synthetic cytokinin which has been recently shown to be an endogenous constituent of anise cell cultures [87]. An avidin-biotin ELISA was developed for benzyladenine measurements in HPLC purified plant extracts by Label and Sotta [182 a)]. The assay utilizes an antiserum raised originally against IPA (*52*), thus, cross-reacting endogenous cytokinins must be removed prior to the analysis of the level of the synthetic cytokinin in the plant tissue.

5.4 Gibberellins

This group of compounds is perhaps the most difficult to analyze. 74 gibberellins have been identified in plants up to now and more than a dozen structures may occur in any given plant, (see: [56]). Different tissues may contain different gibberellins (GAs), and a number of pathways — according to the sequence of hydroxylation steps — can be distinguished for the terminal steps of synthesis of the physiologically active structures [56]. These active GAs occur in exceedingly low levels (in the ng/kg range) in most plant tissues and very little is known about their precise tissue and intracellular distribution, the absolute amounts present, and the physiological changes of their levels. Even for immunological assays, the discrimination of the various GA structures at the physiological level, i.e. against a relatively high background of impurities, represents a formidable task. In addition, only a few GAs, namely GA_3 (*58*), GA_4 (*59*) and GA_7 (*61*) are commercially available (in rather impure form) and the physiologically more important GAs, notably GA_1 (*57*), are not available in large quantities. Some of the interesting structures, such as GA_{19} (*64*) (a key precursor of GA_1 (*57*), or GA_8 (the primary inactivation product of GA_1, = 2β-OH-GA_1) are available in only μg quantities. Thus the set-up of assays for these GAs is precluded for many laboratories. The most conclusive identification of GAs is by GC-MS, but such an analysis is possible in perhaps not more than half a dozen laboratories worldwide — and quantitation of GAs by these techniques has been possible only in a few exceptional cases (see reviews in [56, 330]).

On the other hand, techniques have now been developed to synthesize GA-immunogens using different positions of the GA skeleton, and radiolabeled GAs of high specific activities as well as enzyme-labeled GAs are now available. GA immunoassay is still at an early developmental stage, but the few studies available so far are extremely encouraging. Until ultra-specific antibodies become available, however, some separation step must be included in the protocols prior to the immunological quantification. But even now, for a limited number of GAs, immunological GA analysis already outperforms other techniques in terms of convenience and sensitivity. The earlier GA-sera were described by Fuch's group [102, 103], and these were raised against carboxyl (C7)-coupled GA_3 (*58*). The same coupling site was chosen in some later studies reporting the RIA of GA_3 (*58*) [12, 323, 371], GA_1 (*57*), GA_4 (*59*) and GA_7 (*61*) [12, 331], GA_5 (*60*) and GA_{20} (*63*) [12, 331, 399], and GA_9 (*62*) [12, 256], and the EIA of GA_3 (*58*) and GA_4 (*59*) and GA_7 (*61*) [14]. The latter assays proved particularly sensitive and allowed the detection of less than 1 fmol of the respective GA, per assay. A modified carboxyl-coupling procedure was developed by Yamaguchi et al. [399] who used the N-β-alanyl derivatives of GA_5 (*60*) and GA_{20} (*63*) which were coupled to BSA. The conjugates proved highly immunogenic in rabbits.

From work with the first generation of antisera against gibberellins, it became apparent that C7-linked, unsubstituted GAs, such as GA_9 (*62*) elicited very specific sera with little cross-reaction to other substituted GAs. The substituted more polar structures gave rise to less selective antisera. This can be understood on structural grounds since the bulky substituents of GAs, (e.g. GA_1 (*57*), GA_3 (*58*), GA_4 (*59*), and GA_7 (*61*)) would permit the smaller, unsubstituted GAs, e.g. GA_9 (*62*) to "slip" into the antibody's binding site. The reverse situation, however, would not be possible. This is also the reason why the 1,2-dehydro GAs usually showed more cross-reactions with antisera against the saturated analogues whereas the cross-reaction of the saturated GAs with antisera raised against the 1,2-unsaturated GAs is much smaller, e.g. [12, 14, 399]. All these assays require the GA methyl esters for analysis, which in some cases yields highly apolar structures which are difficult to handle in the aqueous assay environment.

For this and other reason(s), alternative approaches to the synthesis of immunogenic GA protein conjugates were sought. The aim was either to increase assay selectivity or to decrease it in a predictable way, thus allowing the generation of group-selective antibodies. This would allow one to separate and concentrate, from plant material, distinct groups of structurally related GAs prior to physico-chemical analysis. An ideal position to functionalize GAs is the exocyclic double bound common to all GAs. This has been achieved through a number of chemical intermediates [18,181] such as the dihydro GA_{24}-methyl ester-17-cyanoacetate or dihydro-GA_{24}-methyl ester-17-chloroformate [181] to yield highly immunogenic conjugates. In this latter study, the more readily accessible GA_{24} *(65)* was functionalized through both the 17-cyanoacetate and the 17-chloroformate in an attempt to elicit antibodies with high affinities for the related, but very rare structure, GA_{19} *(64)*. Kurogochi et al. [181] found that as expected, the short spacer immunogen, synthesized via reaction of the dihydro-GA_{24}-methyl ester-17-chloroformate with BSA elicited antisera with a lesser degree of discrimination for these molecular sites due to the stronger shielding of the 13-position. Thus, they reported on antisera cross-reacting with GA_{19} *(64)* at 60% relative to GA_{24} *(65)*. These workers were able to quantitate GA_{19} *(64)* in biological specimens (immature seeds of *Pharbitis nil*), by RIA using GA_{24}-methyl ester [methyl-^3H] as the tracer. A one-step route for the synthesis of high specific activity tritiated GA methyl esters was reported and proved applicable for a range of C19 and C20 GAs [181]. It assures high (40%) yields even in the microgram range, and provides a facile access to radiolabeled GA methyl ester suitable for use in conjunction with sera raised against C7-coupled GAs, or coupled GA-methyl esters. In a particularly interesting approach, Eberle et al. used an GA_{13}-19,20-imide hapten-BSA conjugate which mimicked, after coupling, the molecular structure of GA_4 *(59)* to produce monoclonal antibodies reacting with GA_4 *(59)* and a range of related GAs [79]. From their initial hybridoma isolates, highly specific sub-clones recognizing selectively GA_4 *(59)* with very little cross-reactions against other GAs were later derived [Eberle, unpublished]. Monoclonal antibodies against other gibberellins were recently reported [180a]. Whereas most other antibodies require the GA-methyl esters, these latter systems accept the free gibberellins.

Comparisons between RIA and GC-MS proved the reproducibility and accuracy of the immunoassays. This was shown for GA_9 *(62)* analysis in Norway spruce leaf extracts [256], for GA_{19} *(64)* analysis in *P. nil* extracts [181] and for the analysis of GA_5 *(60)* and GA_{20} *(63)* in immature fruits of *P. nil* [331,339]. In all cases, an efficient pre-

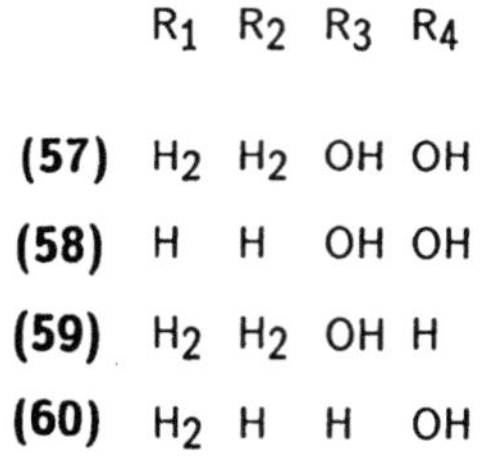

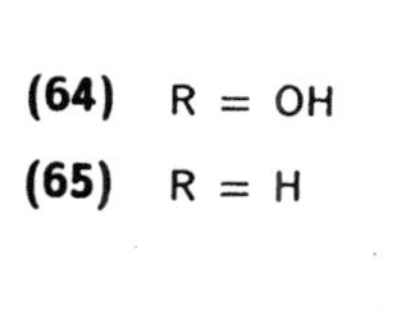

	R1	R2	R3	R4
(57)	H2	H2	OH	OH
(58)	H	H	OH	OH
(59)	H2	H2	OH	H
(60)	H2	H	H	OH
(61)	H	H	OH	H
(62)	H2	H2	H2	H
(63)	H2	H2	H2	OH

(64) R = OH

(65) R = H

cleaning of samples (by HPLC) prior to immunological assay, proved necessary. The assays were used to demonstrate the occurrence of GAs in phloem exudates of tree species [373)], in tissue cultures [117,118,371,378)], and to follow the levels of certain GAs in a variety of physiological responses [13,233)], as well as GAs levels after the application of the synthetic growth retardant, tetcyclacis [117,118)].

5.5 Growth Inhibitors

Antibodies and immunological assays have been developed for the following growth inhibitors: abscisic acid (*48*, ABA), xanthoxin (*66*), and jasmonic acid (*67*). More information is available for ABA (*48*) than any other endogenous growth inhibitor of higher plants. Of the four possible isomers of ABA, 2-cis(+)ABA [= 2(Z)1′(S)ABA] is considered the physiologically active growth regulator [140)]. The R(—) enantiomer has not been found in plants, whereas the 2-trans(+)ABA has been isolated a number of times. 2-trans ABA is in a photochemical equilibrium with the 2-cis isomer and this might explain its occurrence in tissues or extracts exposed to light.

Fuchs and Mayak [105)] were the first to report an immunological assay of ABA (*48*). Anti-ABA antisera were elicited against the synthetic racemate of ABA coupled with hemocyanin via N,N-dicyclohexylcarbodiimide. The assay was a bacteriophage (T4) inhibition assay which, however, required micromolar concentrations of ABA (*48*) and, thus, was too insensitive for most analytical applications.

The introduction of RIA to the analysis of ABA (*48*) [356,380)] lowered the limits of detection to the nanomolar range [356,380)]. In the earlier studies, racemic ABA was used as a starting antigen. However, after synthetic (+)ABA became available [176)], it was found that carboxyl coupling of racemic ABA (*48*) elicited antisera reacting predominantly with the unnatural (—)ABA [232)]. As a consequence, it was necessary to use (+)ABA as starting antigen to allow the setup of precise RIA tests in conjunction with the commercially available, high specific activity tritiated 2-cis(±)-ABA. This was done in a number of studies [232,304,379)]. On the other hand, we found that coupling ABA (*48*) through C_4'-hydrazone intermediates elicited the production of antisera reacting preferentially with (+)ABA (*48*) even if (±)ABA-C_4'-conjugates were used for immunization [381)]. Remaining problems were finally overcome by the selection of hybridomas secreting antibody against 2-cis(+)ABA free acid [180,229,285,286)]. In this case, the cheap cis(±)ABA was used as a starting material and coupled via C_4'-hydrazones. These antibodies [229)] are exceedingly selective and do not recognize (—)ABA, the 2-trans-isomers, or a range of catabolites and potential biosynthetic precursors. The antibodies accept the dissociated form of ABA and are thus ideally suited to react with the compound in aqueous solution at neutral pH. Furthermore, enzyme labeled ABA (coupled via C_4' through a hydrazone) is immunoreactive thus allowing for the set-up of ELISA tests. Later, a number of additional RIA and ELISA procedures for ABA (*48*) were reported [60,63,134,172,194,202,255a,288,304,304a,317,317a,379)], including a second high-specificity MAB against 2-cis(+)-ABA [286a)] for use in RIA certifying that the generation of immunological assays for this polar terpenoid growth regulator is now straightforward, even more so now that pure 2-cis(+)ABA, and tritiated ABA of high specific activity, are commercially available. Commercially available MAB or MAB-based ELISA kits to determine ABA were used successfully in a number of studies, e.g. [72,134c,288,304a,325a].

Reproducibility and accuracy of ABA (*48*) immunoanalysis has been shown in a number of cases, e.g. [172,202,304] and for some tissues, the use of MAB allowed ABA (*48*) determinations in only marginally purified plant extracts [229]. These same monoclonal antibodies were recently used in a micro-scale, enzyme-amplified immunoassay resulting in a large increase in sensitivity. Harris et al. [133] were able to quantitate attomole amounts of ABA (*48*) by enzyme-amplified ELISA in ca. 2 µl assay volumes. Their technique is sensitive enough to allow quantitation of ABA (*48*) in 10 guard-cell pairs (i.e. 20 cells with a total cell volume of 40 pl). This was achieved by dissecting out individual guard cell pairs from lyophilized epidermal tissue. This study showed for the first time the quanitation of a plant growth regulator in a target tissue at the cellular level and is an impressive example of the tremendous potential inherent to immunological assays.

Numerous applications of ABA (*48*) immunoassays have already been reported and these can be mentioned only briefly. ABA (*48*) has been quantitated immunologically in leaves of wheat [172] and other cereals [286a], apple and sweet pepper [304], lupin [286a], broad-bean [229], *Phaseolus* and *Hyoscyamus* [381], tomato, tobacco and Douglas fir [202], and *Citrus* species [134,134a], in epidermal tissue of *Tulipa gesneriana* and *Commelina communis* [289,317], tomato lateral roots [221] roots and leaves of flooded peas [402a], sugar beet tissue [110a,311a] and barley [245a], rice [341a] as well as wheat embryos [245a,288,354a]. The seasonal variation of ABA (*48*) levels in leaves and buds of *Betula papyrifera* and *Acer pseudoplatanus* [381], as well as in Avocado development [25,59,61,394], its occurrence in phloem exudate of trees [373] and tissue cultures such as *Lonicera prolifera* [379] or the kinetics of distribution of the compound in gravireacting stem and root tissue [233] are further examples. Grossmann et al. [117], showed an increase in ABA (*48*) levels of rice tissue cultures after application of the growth retardant, tetcyclacis, which at the concentrations applied (0.1 mM) inhibited sterol biosynthesis in these tissues but had no effect at the levels of the GAs tested. The immunological detection of ABA (*48*) in guard-cell protoplasts of *Vicia faba L.* [367] and isolated guard-cell pairs of *V. faba L.* [133] constituted proof of the occurrence of ABA (*48*) in this cell type which responds to exogenously added hormone by loss of K^+ (this leads to stomatal closure in the intact epidermal tissue). Recently, ABA (*48*) was detected immunologically in extracts from the mammalian brain (mice) and its presence was further confirmed by GC-MS [193].

Much less work has been done with the other growth inhibitors mentioned. Xanthoxin (*66*, XA) is a potential intermediate in ABA (*48*) formation in plants and a growth inhibitor in its own right [334] although newer experiments make it likely that XA (*66*) forms during workup of plant tissue in the light under oxidizing conditions and perhaps does not occur endogenously [140]. Antibodies to violaxanthin were reported in 1979 [198]. Feyerabend and Weiler [95a] have coupled XA (*66*) to proteins via Schiff's bases reduced with borohydride and were able to elicit very strong immune responses with this immunogen in rabbits. The sera proved specific and, e.g., discriminated XA (*66*) from ABA (*48*) but not from violaxanthin and other xanthophylls which had to be removed prior to analysis. Their experiments provided no indication, that, as proposed earlier [315], XA (*66*) redistributes in unlaterally illuminated hypocotyl tissue (sunflowers) during the process of phototropic curvature [95 b)].

The widely distributed senescence promoter, jasmonic acid (*67*) proved to be immunogenic after coupling, through its carboxyl group, to BSA [179]. Knöfel et al. [179] reported a radioimmunoassay for the determination of jasmonate (as the methyl ester) in the 2–200 ng range, in plant extracts of low purity. This assay should be very useful to study in detail, e.g. the occurrence, fluctuations of levels, and taxonomic distribution of this new plant growth regulator.

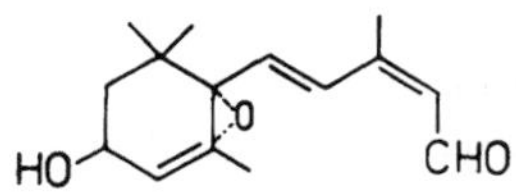

(66)

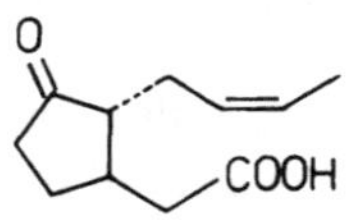

(67)

5.6 Steroids

Brassinosteroids have growth regulating properties in some experimental situations, although their general importance for growth regulation is presently unknown. Horgen et al. reported on monoclonal antibodies elicited against 2α,3α,22α,23α-tetrahydroxy-24β-methyl-B-homo-7-oxa-5-cholestan-6-one (the C24 isomer of brassinolide (*68*)) [149]. Their approach is interesting in that they used a non-covalent steroid-protein adduct prepared by mixing a methanolic solution of the steroid with chloroform/methanol-extracted fetal calf serum at a ratio of 10:1 (w/w). This procedure was originally devised by Landsteiner's group [185], and the adduct is reported to be an effective, but somewhat weak, immunogen [149]. The ELISA based on these MAB was sensitive in the ng range, but apparently of low specificity because all steroids tested (ecdysone, ergosterol, stigmasterol, sitosterol) as well as the natural brassinolide proved highly immunoreactive.

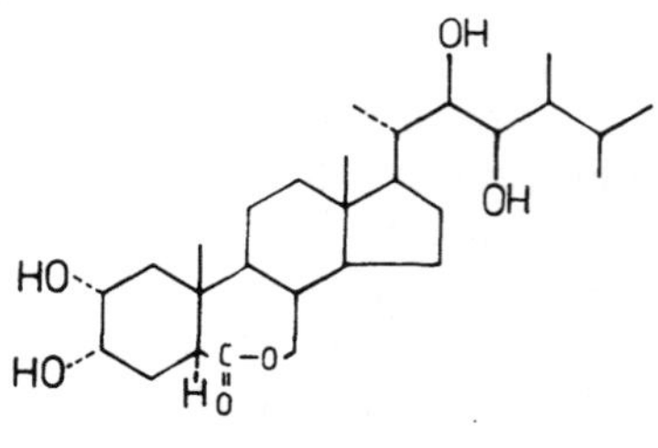

(68)

5.7 Immunoaffinity Chromatography

The use of antibodies to isolate and purify plant growth regulators from plant material or during preparative experiments such as radiosyntheses appears attractive in that the potential of the technique lies in the handling, with minimal losses, of trace components in complex mixtures. Fuch's laboratory was the first to have materialized this potential [104]. These workers prepared affinity purified anti-GA_3 antiserum from rabbits, coupled it to cyanogen bromide activated sepharose and demonstrated that the resulting affinity matrix, which contained about 15 mg of protein bound per ml of gel, retained up to 3 nmol GA_3 (*58*) per ml of gel. The same stationary phase was used in later studies to couple antibody (33% ammonium sulphate cut of crude serum) raised against N^6-benzyladenosine (BA) (column capacity 10 nmol BA/ml gel) [53]. The column also bound kinetin and and isopentenyladenine and thus appeared to be of comparatively low selectivity. Furthermore, both antisera used in these early studies were raised against antigens which do not occur widespread in the plant kingdom.

Significant improvements were made when sera, and later, MAB specifically designed for the analysis of endogenous growth regulators were used for immunoaffinity chromatography. With use of purified anti-IPA antibody coupled to the same support as above, Milstone et al. [234] purified IPA-containing tRNA from total tRNA preparations of various sources and achieved a 12 to 13-fold enrichment.

Systematic studies on the immunoaffinity purification of growth regulators from plant extracts have been carried out in the last few years in the laboratories of Morris [217], Davis [66], Crozier [58 3329] as well as ours [231] and unpublished results.

Ideally, immunoaffinity columns should allow a fast, single-pass, purification with high recoveries of bound ligand and minimum contamination by other substances. Columns should be stable and re-usable. Some of the published systems come quite close to this ideal situation already. Both polyclonal antibodies (from purified sera) and MAB (from purified culture supernatants or ascites fluid) are suitable for this purpose although monoclonal antibodies can be obtained in highly pure form and thus allow maximum column capacities to be achieved. On the other hand, MAB may be rather sensitive even after immobilization [217] and thus yield affinity supports with a short half-life.

In order to achieve maximum enrichment, unspecific adsorption effects to the matrix itself must be minimized. Thus, solid supports based on sepharose [58,104,231,232] agarose [66,231], glycophase-silica [66], CNBr activated microcrystalline cellulose [217] or glutaraldehyde-activated silica [329] appear to be preferable to vinyl-based stationary phases such as Fractogel [231]. Because of its excellent flow rates, high protein-binding capacities, retention of antibody function and minimum unspecific adsorption, Affigel-10 is perhaps the preferable support for plant growth regulator affinity chromatography [66,231] although CNBr activated cellulose [217] or Sepharose also performed well in several studies [104,180,217,232]. The latter support gives about a two-fold lower final antigen-binding capacity [66,231]. With Affigel-10, we have consistently achieved binding capacities between 2 and 8 nmol of growth regulator per mg of immobilized protein (50% ammonium sulphate cuts of various ascites fluids). Elution of bound hormones from affinity columns may require chaotropic treatment [180] or acidic pH [180], or solvent elution with methanol [58,217] or even acetone [232]. We have reused anti-ABA columns [232] more than 25 times in an acetone/buffer cycle without apparent loss of binding capacity and similar observations were reported by Crozier et al. [58], however, column stability may vary greatly from antibody to antibody (R. Morris, personal communication). Davis et al. [66] have compared several solid supports in conjunction with polyclonal antibodies (purified sera) against (+)ABA (*48*), zeatin riboside (*54*) and dihydrozeatinriboside (*56*). They were able to isolate, after a single pass, analytically clean (HPLC with base-line separation) cytokinins from soybean seeds and (+)ABA (*48*) from soybean leaves [66]. Similarly, Sundberg et al. [329] achieved a dramatic purification of IAA (*49*) from *Pinus sylvestris* shoot extracts in a single run using polyclonal antibodies purified from sera. Similarly, indole-3-acetamide was purified from *Nicotiana tabacum* crown gall tissue by single-pass immunoaffinity chromatography on divinylsulfone-activated agarose to which MAB against IAA (*49*) were coupled [342 a]. MacDonald and Morris have recently reviewed [217] their techniques to isolate cytokinins by immunoaffinity chromatography. They have also reported very high enrichment of several cytokinins from crown gall extracts or *Agrobacterium tumefaciens* cultures after a single-pass chromatography allowing direct quantitation after HPLC separation. Their system has been partially automated [400].

Careful evaluation of the immunoaffinity separation is required because antibodies, per definition, have the tendency to co-purify closely related structures (which may also otherwise be difficult to resolve). E.g. Crozier et al. have detected an impurity in immunoaffinity purified IAA (*49*) from *Zea mays* tissue only after successive HPLC passages of the eluate [58]. An interesting application of immobilized, stereoselective antibodies, which may have more widespread applications, is the resolution of racemic, tritiated ABA of high specific activity. This procedure resulted in greater than 95% optical purities of both enantiomers, in a single pass, using Sepharose-bound polyclonal antibodies against (+)ABA and (—)ABA [232]. A similar technique, based on monoclonal antibodies was recently also reported [180].

5.8 Immunocytochemical Localization

Only a few immunocytochemical studies attempting the determination of intraorgan or even intra-cellular distribution of plant hormones have been done. Zavala and Brandon [401] used F_{ab} fragments of gold or rhodamine conjugated immunoglobulins from rabbit antisera against DHZR (*56*) [26] to localize DHZR in sections of roots at the light microscopic level [401]. Sotta et al. [322] employed rabbit anti-(±)ABA sera and Sternberger's PAP technique [328] to detect ABA (*48*) in *Chenopodium polyspermum* tissue. Prior to the immunological reactions, the tissue was rapidly immersed in 2% aqueous, water-soluble carbodiimide (EDC, (1-(3-dimethylaminopropyl)-3-ethylcarbodiimide · HCl). It was then assumed that protein-linked ABA was detected by the immune serum (which was raised against C_1-coupled ABA, synthesized by a similar, carbodiimide-mediated reaction). Sossountzov et al., from the same laboratory [321], have extended these initial studies to the immunoelectron microscopic localization of ABA (*48*) with colloidal gold in Lowicryl-embedded, EDC treated tissue sections of *Chenopodium polyspermum*. They were able to demonstrate the presence of immunoreactive material in the cytoplasm and chloroplasts of cells of the nodal inner cortex. Eberle et al. [78,80] were able to localize, by immunoelectron microscopy using the immunogold technique and monoclonal anti-isopentenyladenosine antibodies, and low temperature embedding in Lowicryl K4M, isopentenyladenine-like immunoreactivity in the cell wall of certain protonemal cells of cytokinin overproducing mutants [357] of the moss, *Physcomitrella patens*.

In this study, no attempt was made to chemically couple the cytokinin to macromolecular cell components prior to tissue embedding. Rather, it was attempted to reduce antigen losses by embedding at sub-zero temperatures. Chemical fixation of cytokinins was reported by Sossountzov et al. [321 a)] to occur during treatment of tissue with sodium periodate, followed by borohydride reduction and fixation of the tissue with paraformaldehyde/glutaraldehyde mixtures. A range of controls was used to show the specificity of the reaction which would immobilize the cytokinin-ribosides as well as other cytokinin-glycosides. Of these, presumably only the cytokinins-ribosides would be detected by the two antibodies used. The technique was used to study, at tissue/organ level, the distribution of immunoreactive cytokinins (IPA and tZR) in two isogenic tomato lines, Craigella and the sideshootless mutant, Craigella lateral suppressor.

Thus, it is technically possible to demonstrate rather clear immune reactions for growth regulators on tissue sections of plant specimens, even at the ultrastructural level. Although by a number of standard controls included in all the available reports, the reactions observed appear to be clearly distinguishable from nonspecific effects, the immunolocalization of small molecules is still confronted with an array of obstacles. This limits the conclusions which can be drawn from such experiments: a) it is not possible to accurately assess workup losses of the antigen during tissue preparation and thus, uneven losses from different areas of the specimen, or from different compartments may go undetected. E.g., the degree of immobilization obtained with the chemical fixation techniques targeting to proteins [321,322)] should greatly depend upon the protein concentration of the cells/compartments. Consequently, it is very hard to draw quantitative, or even semi-quantitative conclusions from immunohistochemical specimens. Perhaps gas-phase chemical immobilization of antigens in the specimens offers a solution to this problem. In addition, physical immobilization might be possible in some instances. E.g. photochemical activation of natural photoaffinity labels such as ABA (*48*) [150)] might prove effective in linking the regulator to its receptor. The immobilized hormone could then be detected with specific antibodies in suitably processed specimens.

b) Plant tissues often contain high levels of structurally related compounds, or growth regulator conjugates, even in macromolecular form, which might preferentially label. This is not easy to detect with the usual controls carried out in such experiments. In most studies, the authors have therefore sought additional indications to validate their conclusions. Zavala and Brandon [401)] have avoided an interference by cytokinin-containing tRNAs by using DHZR antibodies. DHZR (*56*) does not occur in tRNA. Sossountzov et al. [321)] used a water-stress program which is known to increase the level of endogenous ABA (*48*) drastically (but also that of conjugates) as an additional control and Eberle et al. [78,80)] compared cytokinin overproducing mutants of *Physcomitrella* with wild-type moss tissue. Model systems, such as the one described by Skene et al. [317 b)] for ABA (*48*) will be very useful to estimate the sensitivity of immunolocalization, the efficiency of chemical immobilization procedures as well as losses during tissue preparation, of LMC.

6 Immunoassays for Pesticides and Pesticide Metabolites

6.1 Scope and Potential

Pesticides and their metabolites are usually analyzed by physico-chemical techniques such as GC, HPLC or GC-MS. The protocols employed are often complex and most of these techniques are cumbersome, slow, and consequently expensive.

For a range of applications, physico-chemical analysis is indispensable (e.g. for structural confirmation, the GC-MS technique), however, there are other areas which will benefit from the application of immunoassays. These are equally sensitive (or more sensitive), involve less complex protocols and are high-capacity batch techniques. Thus, they are faster and considerably less expensive than physico-chemical analysis.

Virtually all stages of development of a commercial pesticide as well as its later application require the availability of suitable analytical techniques applicable to a wide range of samples such as animal, plant, soil, water, or human body fluid samples. The current trends are towards decreasing levels of application in the field, producing more effective pesticides and increasing the monitoring of metabolic derivatives of the applied compounds. Thus pressure continues to built towards analytical procedures of ever increasing sensitivity. From the foregoing chapters, it has become clear that immunological assays can match even extreme needs in this respect.

There is also increasing political and social pressure towards maximizing safety testing of chemicals released into the environment and food chain. Such demands could more easily be met if fast, inexpensive and high-capacity, yet reliable, techniques were at hand. The development of a new pesticide is a process of several years. The development of a high-performance immunoassay, based on the inexhaustible reservoir of MAB may take, for a specialized laboratory, approximately a year in complicated cases. Thus, although immunoassay set-up initially may be slow and also cumbersome, an integration of assay development into the overall process of developing a new pesticide would pose no problems. We see the following applications of immunoassays for pesticides:

a) determination of distribution, transport, decomposition and/or excretion of the pesticide in the treated organism; residue analysis;
b) monitoring of key degradation products, their distribution, transport etc. in treated organisms; metabolic studies;
c) monitoring of soil, water or aereal levels of pesticides and/or metabolits after application (e.g. spray drifts, ground water levels, distribution and persistence of pesticides in the soil); environmental screening in the widest sense;
d) monitoring of pesticide intake, e.g. by farm workers, monitoring of pesticide levels in clinical and epidemiologic studies, determinations of pesticide levels in food products;
e) optimization of pesticide application programs to ensure most economical use of pesticides in the field.

Since the first report on the production of antisera against pesticides in 1968 [121)], a significant advance has been made in this field and the current state-of-the-art will be reviewed in this chapter. Reviews on part of this topic have appeared else-

Table 5. Immunoassays for the determination of insecticides. The following abbreviations are used: — = data not available; C = carbon-14, all others as in Table 4. If not stated otherwise, data are for standards

Compound	Assay format	Detection limit (D) Measuring range (M)	Reproducibility	Applications	Refs.
Diflubenzuron (*69*)	E,P,SP,AP	D 1—40 ppb M 10—200 ng/ml	W < 4	water, milk	385
	E,P,SP,AP	D 2 ppb (milk) M 2—200 ng/ml (milk)	—	milk	386
Paraoxon (*70*)	R,P,LP,T	D 1 pmol/assay M 0.7—70 pmol/assay	—	—	138 a
	E,M,SP,AP	D 1 µg/ml M 10^{-5}—5×10^{-3} M	W 8—14	—	28
Parathion (*71*)	R,P,LP,T,C	D 4—20 ng/assay M 10—500 ng/assay	—	fortified human plasma and lettuce	83
s-Bioallethrin (*73*)	R,P,LP,T	M 0.1—10 nmol/assay	—	—	393
	E,P,SP	M 0.0016—1.6 nmol/assay	—	milk	346 b
Dieldrin (*74*)	R,P,LP,I	D 150 pg M 1—100 pmol/assay	—	—	188
	E,P,LP,GD	D < 10 ng/assay	—	—	94 a

where [129,346,392]. The reader is especially referred to the recent excellent reviews on the utility of immunoassay in pesticide trace analysis [129 a] and on immunological assays for agrochemicals [245 b]. These articles also provide critical accounts of the advantages and disadvantages of applying immunoassays to problems of pesticide analysis. These discussions will therefore not be repeated here. It appears, though, that especially now that the hybridoma technology for the production of anti-hapten MABs is fairly well established, e.g. [113,361,375], much more benefit could be drawn from an increased use of immunoassays in pesticide analysis.

6.2 Insecticides

In terms of immunological behaviour, perhaps the best characterized insecticides are diflubenzuron (*69*), studied by Hammock's group [385,386,387], and the organophosphorous compounds, paraoxon (*70*) [28] and parathion (*71*) [343], studied by Hunter and colleagues and Mumma's group, respectively.

The chitin synthesis inhibitor, diflubenzuron (*69*) (N-[[4-chlorophenyl)amino]-carbonyl]-2,6-difluorobenzamide) and the related BAY SIR 8514 (*72*) (N-[[[4-(trifluoromethoxy)phenyl]amino]carbonyl]-2-chloro-benzamide) were functionalized by reacting an appropiately substituted aniline with 2,6-difluoro- or 2-chlorobenzyl isothiocyanate. This introduced free carboxylic acid groups which were subsequently used to couple to amino groups of carrier proteins via active ester intermediates or the water-soluble carbodiimide method. The following conjugates were synthesized:

i) N-(carboxypropyl)diflubenzuron-KLH, ii) N-(carboxypropyl)diflubenzuron-BSA, iii) N-(carboxypropyl)diflubenzuron-KLH and iiii), diflubenzuronphenoxyacetic acid-KLH [387]. All conjugates (with 2–5 mol of hapten/50.000 daltons of protein) proved immunogenic in rabbits although i) yielded low titers. Sera from animals immunized with either of these immunogens were then cross tested in an indirect ELISA for reversible binding to various hapten-ovalbumin solid phases (e.g. polystyrene adsorbed ovalbumin conjugates). This allowed the influence of "bridge-effects" of the haptens and conjugates to be assessed. As could be expected for polyclonal antibodies, the combination of serum with the appropriate immobilized antigen was of critical importance. The optimum combination was found to be antiserum raised against N-(carboxypropyl)diflubenzuron-BSA with immobilized diflubenzuron-phenylacetate-OA, i.e., a situation in which both the chemical nature of the bridge and the position of linkage were different [385]. Specificity checks revealed a marked inhibition of binding of antibody to the immobilized antigen by diflubenzuron (*69*) and BAY SIR 8514 (*72*), but 2,6-difluorobenzamide and 4-chloroaniline hemisuccinamide were ineffective competitors. However, sera were found which also sharply discriminated diflubenzuron (*69*) from BAY SIR 8514 (*72*) showing the ability of antibodies to distinguish both substitution of the benzamide portion, as well as of the aniline moiety [385]. By ELISA, diflubenzuron (*69*) could be measured in the 10–200 ng ml^{-1} range. Compared to classical HPLC and GLC techniques, ELISA required smaller sample sizes, shorter assay times, offered higher assay capacities per man-day and, since less steps were involved in the overall procedure, was more reproducible (coefficients of variation: 3.5%). In a later study [386] the system was reinvestigated and further optimized by using BAY SIR 8514-immunogens, as well as an additional type of linkage made by coupling the diazonium derivative of 4′-amino-diflubenzuron to BSA. By careful optimization of the ELISA system and appropriate choice of the serum/immobilized antigen combinations, the authors were able to determine diflubenzuron (*69*), BAY SIR 8514 (*72*) as well as penflurone, in the 100 ng ml^{-1} range. Diflubenzuron (*69*) could be detected in the ppb range in 10 ml samples of unprocessed milk [381].

Ergovich et al. [83] coupled reduced parathion (*71*) (O,O-diethyl O-(*p*-aminophenyl thiophosphate) though a diazonium link, to BSA and were able to raise anti-

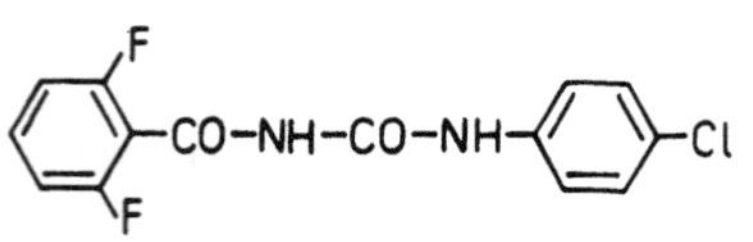

(69)

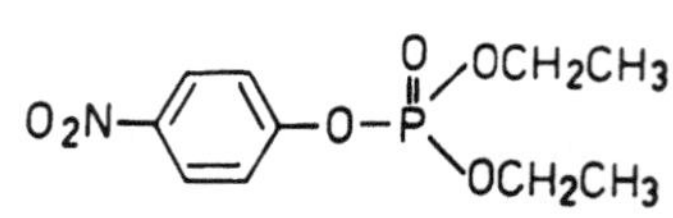

(70)

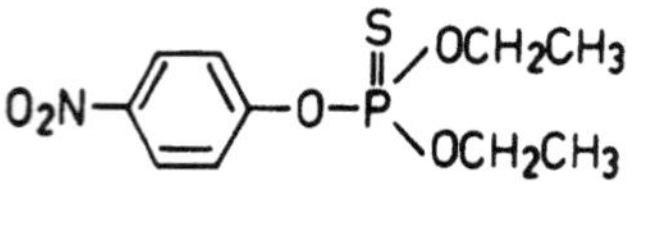

(71)

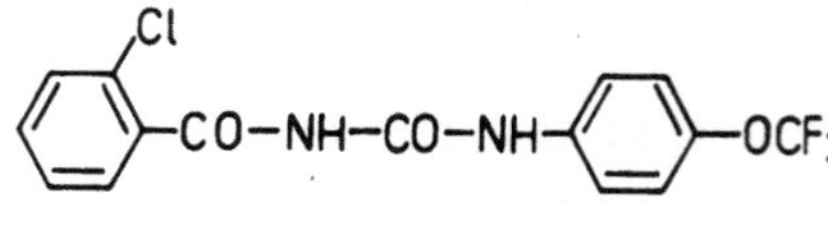

(72)

sera recognizing parathion (*71*) and discriminating it clearly against similar compounds such as paraoxon (*70*). They used low specific activity [^{14}C]parathion as tracer in RIA to detect, in ng amounts parathion in human plasma and lettuce extracts. The detection limit was given as 10—20 ng of parathion in crude samples.

Vallejo et al. [343)] synthesized various immunogens containing parathion-derivatives as haptenic groups. 2-Aminoparathion was linked to BSA using glutaraldehyde cross-linking or the hemisuccinate. 2-Nitro-5-hydroxycinnamic acid was converted to the corresponding 3-substituted parathion by reaction with diethyl thiophosphorylchloride and the carboxylate product was coupled to BSA. In addition, reduced parathion was coupled through the amino function using glutardialdehyde or after conversion to the 4-nitro-4′-benzanilide, to BSA via a diazonium linkage. The immunological properties of the 5 conjugates were examined in rabbits. It was found that although antibodies against the hapten structures were readily detectable, very little binding of underivatized [^{3}H]-paraoxon was observed, indicating that the structural changes introduced during immunogen synthesis were too severe. Paraoxon (*70*), in addition, has an insufficient size to fill an antibody's antigen binding site completely. In such cases, some contributions of the "bridge" to the antigenic epitope must be expected. The original approach of Ercegovich et al. [83)] to couple reduced parathion (*71*), directly, without a spacer, appeared to be a more effective way to elicit an appropriate immune response. Monoclonal antibodies obtained, using diazo linked (*p*-aminophenyl)-paraoxon and KLH as carrier protein [28)], have the ability to discriminate paraoxon (*70*) from, e.g., parathion (*71*), although the latter retained a ca. 3% cross-reaction relative to paraoxon (*70*). However, the MAB were of rather low affinity. This, although resulting in a wide assay range, severely reduced assay sensitivity. Sensitivity was found, in an ELISA, to be ca. 1 ng ml^{-1} paraoxon (*70*) and the measuring range typically extended from 10–100 ng ml^{-1} of the insecticide. Antisera against paraoxon (*70*) were reported earlier from the same group [156)]. A quite different approach was used by Heldman et al. [138 a)] who coupled paraoxon (*70*) through the phosphorous group to BSA. They reacted O-(4-nitrophenyl)-O-ethyl phosphorochloridate, prepared from the corresponding amine, either directly with BSA, or with β-alanine first, followed by activation through the N-hydroxysuccinimide ester and then coupling of the active ester to BSA [138 a)]. Antisera were raised in rabbits and screened for their ability to prevent paraoxon-elicited inhibition of acetylcholine esterase, as well as for their suitability for RIA. RIA detected about 300 pg (~ 1 pmol) of paraoxon (*70*). The antibodies recognized, as expected from the immunogen structures, the structurally related parathion (*71*) nearly as well as paraoxon (*70*). This example once again shows the influence of the antigen coupling site on antibody selectivity (cf. with results in Ref. [28)]).

A model substance for organophosphorous agents, methyl phosphonic acid, *p*-aminophenyl 1,2,2-trimethylpropyl diester, MATP (*77a*) was diazotized, coupled to HSA and administered to chicken as well as rabbits [310 a)]. A variable immune response was obtained in the two animal species with some individuals developing significant titers. Competitive inhibition EIAs (ELISAs) were set up by immobilizing MATP, through Schiff's bases, directly on glutardialdehyde-activated polystyrene microtitration plates. The technique detects between 10^{-8} and 10^{-4} M standard MATP with either the rabbit or chicken antibodies. The assays were so far not applied to biological samples.

The pyrethroid insecticide, S-bioallethrin (*73*) (1R,3R,4'S isomer) was determined in the pmol range, by RIA using a stereoselective antiserum raised against S-bioallethrin-hemisuccinoyl-conjugates [391,393]. The antiserum displayed significant affinities for R-bioallethrin (1R,3R,4'R), but reacted significantly weaker with e.g. the 1S,3S,4'S; 1R,3S,4'R or 1S,3S,4'R form and the 1R,3S,4'S and 1S,3R,4'R isomers were almost unreactive. This shows the ability of antibodies to discriminate between optical isomers. This is an especially useful feature of immunoassays because, by physico-chemical trace analysis, the selective determination of chiral compounds is nearly impossible.

A recenty described ELISA for S-bioallethrin (*73*) was compared to a RIA and found to be more sensitive and precise [346 b]. Whole milk was analyzed for the presence of the insecticide, and the detection limit was given as 1 μg S-bioallethrin (*73*) per ml of 50% milk.

Wraith et al. [397 a] have attempted to develop immunoassay techniques for another pyrethroid insecticide, cypermethrin (*73a*) and at the same time for the plant metabolites, 3-phenoxybenzoic acid and dichlorovinyl cyclopropane carboxylic acid by coupling the two metabolites, rather than cypermethrin itself, to carrier proteins. They have successfully raised antisera gainst the two haptens and developed ELISA techniques for the metabolites. The assay for dichlorovinyl cyclopropane carboxylic acid is also sensitive to cypermethrin (*73a*) and permethrin whereas the assay for 3-phenoxybenzoic acid can be used to analyze a range of synthetic pyrethroids containing this moiety. The assays were applied to the analysis of tea samples.

Similarly, an antiserum against the fenvalerate (*73b*) metabolite, 2-(chlorophenyl)-isovaleric acid which was linked to BSA through a γ-aminobutyrate bridge, was raised and a RIA developed which detects ca. 2 ng of 2-(chlorophenyl)isovaleric acid in urine samples [257 a].

Radioimmunoassays for the organochloric insecticides, dieldrin (*74*) and aldrin (*75*) were described by Langone and van Vunakis [188]. These authors coupled 6,7-dihydro-6-carboxyaldrin, as the activated N-hydroxy-succinimidester, to HSA. Their assay detects picomoles of dieldrin (*74*) and aldrin (*75*), but also reacts with structurally related insecticides, e.g. endrin, heptachlor and chlordane. An 18% cross-reaction is reported for DDT (*76*) and 19% for decachlorobiphenyl. It appears, that this assay system is mostly selective for the hexachloro-bicyclic ring system and some sort of samples separation will be required prior to the analysis of dieldrin (*74*) and aldrin (*75*) in order to exclude related structures. A homogenous immunoassay (EMIT®) for dieldrin (*74*) was recently reported in abstract form [94 a].

Centano et al. [44] have successfully raised antisera against DDT (*76*) (via 2,2-bis[4-chlorophenyl]-acetic acid protein conjugates) and malathion (*77*) via the anhydride of di-deethylmalathion, coupled to protein. They reported on a hemagglutinin inhibition reaction to detect these antigens and this was one of the earliest reports [44,121] on pesticide immunoanalysis. Table 5 lists a selection of assays according to the assay formats used and gives the most important assay parameters.

6.3 Fungicides

Immunological assays have been reported for the systemic fungicides, triadimefon (*78*) [248], metalaxyl (*79*) [249] and benomyl (*80*) [247] for the benomyl metabolites, 2-

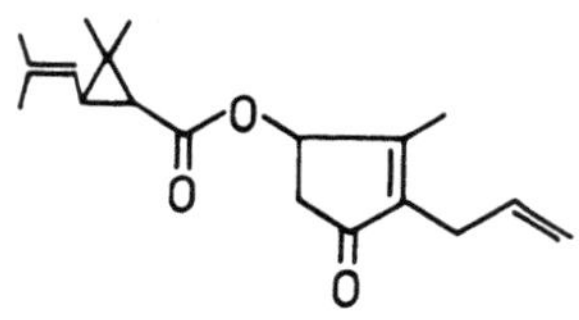

(73)

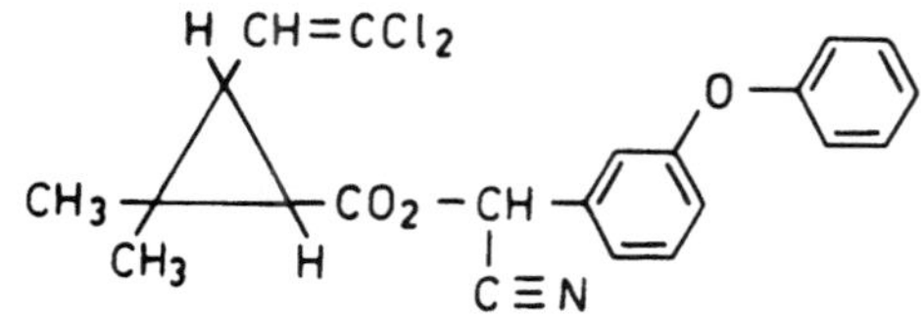

(73a)

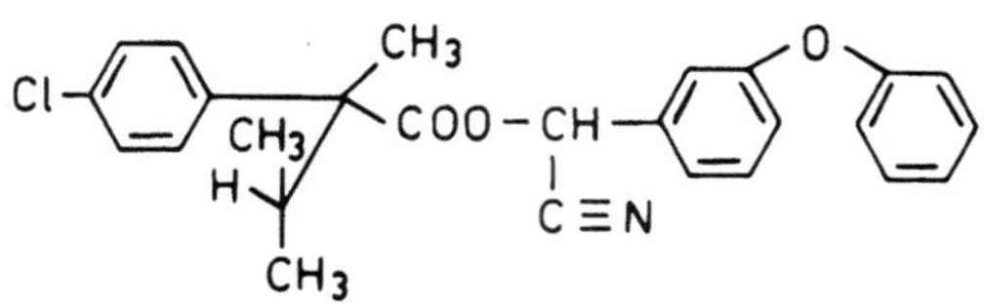

(73b)

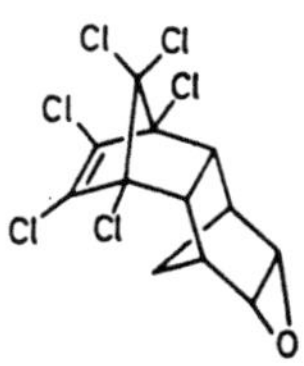

(74)

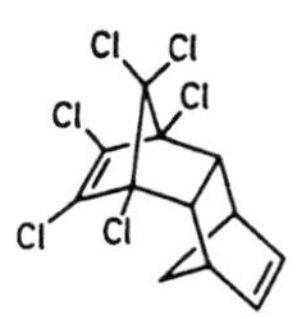

(75)

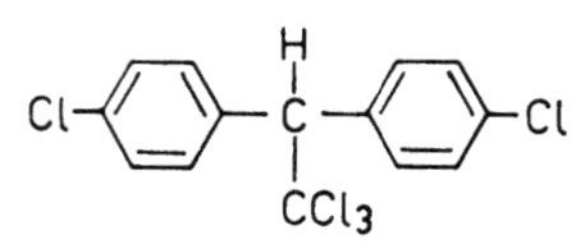

(76)

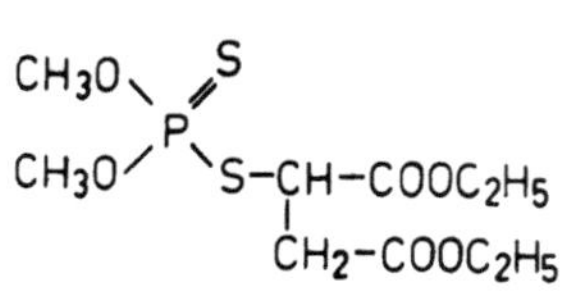

(77)

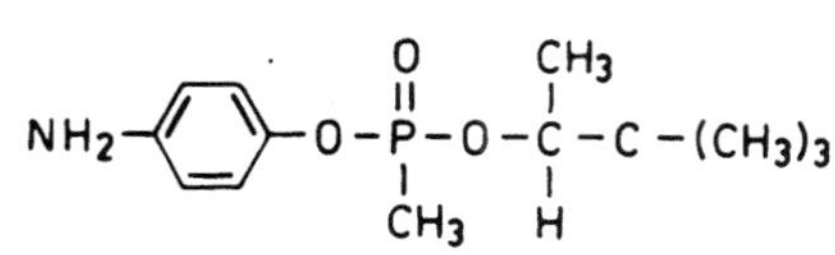

(77a)

aminobenzimidazole (*81*) and methyl-2-benzimidazolycarbamate (*82*)[247] and for the contact fungicide, iprodione (*82a*)[247a].

Newsome and Shields[247] made use of the fact that benomyl (*80*) (methyl 1-(butylcarbamoyl)-2-benzimidazole-carbamate) converts quantitatively to MBC (methyl 2-benzimidazolecarbamate) when heated in ethyl acetate, to determine, by RIA, benomyl plus MBC, after conversion of the benomyl (*80*) to the MBC. The assay used an antiserum raised in rabbits against 2-succinamidobenzimidazole coupled to HSA. The assay detected 1–2 ng of MBC and was used for MBC analysis in several vegetable and fruit tissues. 2-Benzimidazolylurea effectively cross-reacted with MBC, but 2-aminobenzimidazole, a minor benomyl metabolite, reacted more weakly (ca. 8% relative to MBC) and is not likely to interfere in MBC residue analysis. Antisera against 2-aminobenzimidazole were also described[212].

Table 6. Immunoassays for the determination of fungicides. For abbreviations, see Table 4 and 5. If not stated otherwise, data are for standards

Compound	Assay format	Detection limit Measuring range	Repro-ducibility	Applications	Refs.
Triadimefon (*78*)	E,P,SP,PO	M 1—16 ng/ml	W 3 W 5(pear)	fortified commodities e.g. pear, apple, grape	248
Metalaxyl (*79*)	E,P,SP,PO	D 63 pg/ml M 0.07—1 ng/ml	W < 5	fortified commodities, e.g. potato, tomato squash	249
Benomyl (*80*) and methyl-2-benzimid-azolylcarbamate (*82*)	R,P,LP,C	M 1—20 ng/assay	—	fortified commodities, e.g. orange, grapefruit	247
Iprodione (*82a*)	E,P,SP,PO	D 0.03 mg/kg M 0.2—12 ng/ml	W 7.8 B 9.9	fortified commodities, e.g. tomato, grape	247a

Newsome [249] acid hydrolyzed metalaxyl (*79*) (methyl N-(2-methoxyacetyl)-N-(2,6-xylyl)-alaninate) to the corresponding dicarboxylic acid then coupled this to HSA with EDC, and to ovalbumin with the mixed anhydride technique. The HSA conjugate was used as immunogen and the ovalbumin conjugate was used as a coating antigen for ELISA. Straight line standard curves were obtained from ca. 0.05–1 ng ml^{-1} metalaxyl (*79*). Although several related structures, such as metolachlor, diethathyl ethyl or furalaxyl cross react significantly, very weak cross-reactions (less than 0.1%) were observed with propachlor or benzoylprop-ethyl. Metalaxyl (*79*) fortified commodities such as cucumber, squash, avocado, tomato and potato samples, when analysed by ELISA as well as a standard GC technique, revealed the accuracy of the ELISA determination. However, the presence of one of the cross-reacting compounds, if undetected, would probably lead to false positive identifications.

Very recently Newsome reported on the analysis of triadimefon (*78*), by ELISA [248] using a rabbit antiserum raised against succinyl-tradimenol-HSA. Correlation analysis of spiked (0.5–2 ppm) apple, pear, pineapple and grape samples with ELISA and GC gave comparable precision (ca. 5%) and proved the accuracy of the ELISA technique. A RIA method for the analysis of ethylcarbamate in urine has also been described [390]. The contact fungicide, iprodione (3-(3,5-dichlorophenyl)-N-isopropyl-2,4-dioxoimidazolidine-1-carboxamide) (*82a*) is not readily linked to carrier proteins. Newsome [247a, 247c] used the 2,4-dioxoimidazolidine-1-propionic acid analogue of iprodione for linkage to HSA. Sera were raised in rabbits and then used in an ELISA to determine iprodione (*82a*) in the ng/ml range in various commodities (tomato, cucumber, grape, strawberry) to which the fungicide had been added in the 0.1–1 mg/kg range: 86–91% of it were recovered by the ELISA technique (78–92% by the reference GC technique). Several other fungicides, such as vinclosolin and procymidone, as well as several iprodione derivatives are detected by the assay, vinclosolin and procymidone at a significantly lower level than iprodione itself. This restricts the use of the assay to such samples in which the occurrence of these other fungicides can be excluded.

Table 6 compiles the assay formats and basic assay parameters of representative techniques discussed in this chapter.

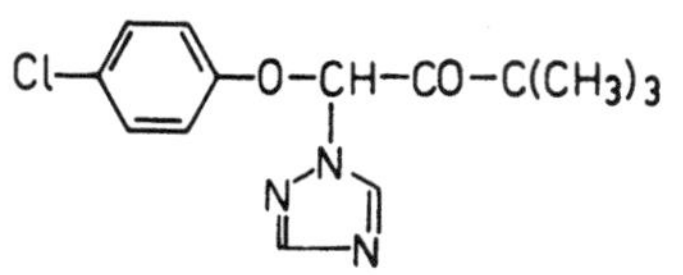

(78)

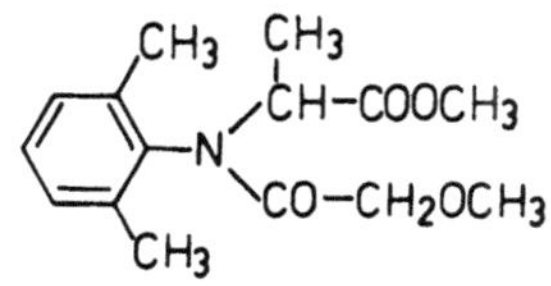

(79)

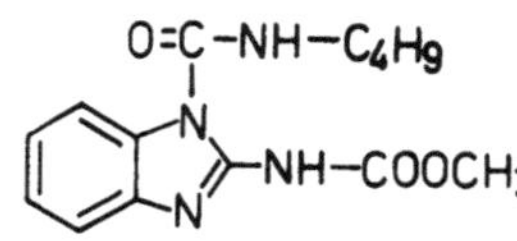

(80)

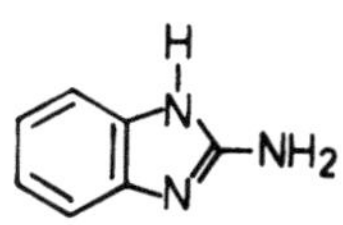

(81)

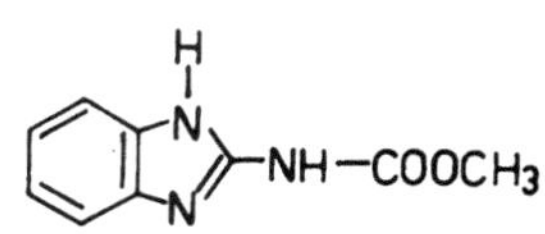

(82)

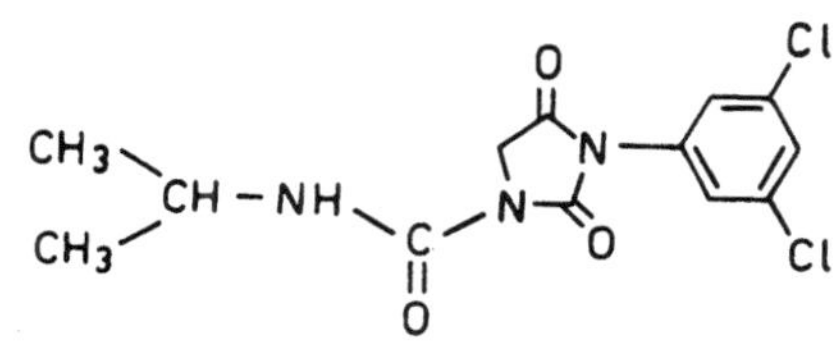

(82a)

6.4 Herbicides

Radioimmunoassays for the analysis of the auxin herbicides 2,4-D (*50*) and 2,4,5-T were discussed in Chapt. 4.2.

For the analysis of paraquat (*83*), an extensively used herbicide, immunoassays were developed for clinical application [54a,204,253,254] as well as for the analysis of field-generated samples [345] and food products [346a]. Immunogen synthesis was effected linking N-(4-carboxybut-1-yl)-N′-methyl bipyridilium [93,345] or N-(4-carboxyprop-1-yl)-N′-methyl bipyridilium [345] to various proteins such as BSA [93,345], conalbium, OA, HSA, or KLH [345]. The KLH-conjugate was found to be the best coating antigen while the conalbium conjugate gave best immune responses [345]. The standard curve range of ELISA was from ca. 0.1 to 100 ng/ml paraquat (*83*). As expected, the methyl-propyl analogue of paraquat was equally reactive whereas paraquat metabolites and related compounds such as diquat (*84*) or difenzoquat gave very little cross reaction. In a worker exposure study, paraquat (*83*) was analysed by ELISA and GC, with comparable results. The assay was also evaluated with body fluids, such as plasma, lymph and urine [345]. This work was recently extended to paraquat residue analysis, by ELISA, in milk, beef and potatoes [346a]. The synthesis of a fluorescent paraquat derivative and its use in a sensitive fluorescence immunoassay (FIA) for paraquat (*83*) in plasma was also recently reported [54a]. FIA was employed by Lukens et al. [213] to analyze diquat (*84*), hexachlorophene, and diethylstilbestrol (*95*). These authors used ionizing radiation to develop complex antigen-carrier protein conjugates.

Diclofop-methyl (*85*) (methyl-2-[4-(2,4-dichlorophenoxy)phenoxy]propionate), is a selective herbicide for the control of weed grasses. Its synthetic precursor, diclofop acid, was coupled via the carbodiimide method, to BSA [312]. Sufficiently high-titer

Table 7. Immunoassays for the determination of herbicides. The following abbreviations are used: FL, fluorescein; all others as in Tables 4 and 5. If not stated otherwise, data are for standards

Compound	Assay format	Detection limit Measuring range	Repro- ducibility	Applications	Refs.
Paraquat (*83*)	E,P,SP,AP	D 0.1 ng/ml M 0.1—27 ng/ml	W < 4 B < 5	worker exposure	345
	R,P,LP,I,T	D 0.1—0.25 ng/ml M 10—2500 ng/ml	—	clinical	93
	F,P,LP,FL	D 20 µg/l M 50—2000 µg/l	W 6—9 B 7—11	clinical	54a
	E,P,SP	D < 1 ppb (milk) M 0.1—30 ng/ml	—	fortified commodities, e.g. milk, potato, beef	346a
Diclofop-methyl (*85*)	(a) E,P,SP,PO	D 23 ng/ml M 10—70 ng/ml	—	fortified extracts, e.g. soil, milk, soybean	312
	(b) F,P,LP,FL	D 45 ng/ml M 20—200 ng/ml	—		
Terbutryn (*86*)	E,P,SP,AP	D 25 µg/l (water) M 20—1200 pmol/assay	—	surface water	153
Atrazine (*86a*)	E,P,SP,AP	D 0.1 ppt M 1.1—2200 ppb (water)	—	surface water	152a
Molinate (*86c*)	E,P,SP,AP	D 3 ng/ml M 20—2000 ng/ml	W 7.8 B 9.4	surface water	316b

sera were raised in rabbits which could be used in ELISA and FIA. To this latter end, diclofop acid was conjugated to fluorescein amine by the mixed anhydride technique. Both FIA and ELISA gave comparable assay sensitivity with detection limits of 23 and 45 ng/ml respectively. The antisera bound both enantiomers of diclofop-methyl (*85*) equally well and showed cross reactions with 2-methyl-1-methoxy-2-oxoethyl-2[4-(2,4-dichlorophenoxy)phenoxy]propionate as well as 2-[4-(2,4-dichlorophenoxy)-phenoxy]propanoic acid. However, 2,4-D (*50*) [348], 2,4,5-T or MCPA ((4-chloro-2-methylphenoxy)acetic acid) showed no cross-reactivity. Both FIA and ELISA gave recoveries consistently around 97–102% at various spike levels (ca. 40–400 µg ml^{-1}). The detection limits for diclofop-methyl (*85*) analysis for a number of natural samples is in the range of 0.023—4.6 ppm. ELISA as well as FIA was shown to be accurate when compared to a standard GC/LSC assay [312]. The S-triazine herbicide, terbutryn (*86*) was convented to the sulfoxide derivative by oxidation with *m*-chloroperbenzoic acid in chloroform [153]. The resulting terbutryn sulfoxide was coupled to BSA (immunogen) and alkaline phosphatase (enzymer tracer). ELISA gave linear dose-response curves (logit/log) in the range of 20 to 1200 pmol terbutryn/assay. Cross-reactants were ametryn and atrazin, whereas simetryn, simazine, prometryn and propaxine were only marginally or not all cross reactive. Thus, this assay permits the determination of terbutryn (*86*) as well as its hydroxy-metabolite, after separation from each other and, if required, other immunoreactive triazines [153]. The same group reported on the analysis of atrazine (*86a*) by immunoassay [153c] which was used to analyze the herbicide in surface water samples [153b]. The serum also allowed to immunolocalize atrazin conjugates in leaves of maize [153a]. The same group raised sera against atrazine (*86a*) by converting the analogous ametryn (2-ethylamino-4-

methylthio-6-(2-propylamino)-1,3,5-triazine) to the corresponding sulfoxide followed by coupling to hemocyanin [152a)]. ELISA and the use of an affinity purified serum allowed to determine from 10 fmol to 30 pmol of atrazine (*86a*) per assay. A range of S-triazines, including ametryn, prometryn, simetryn, propazine and simazine cross reacted in the assay. Terbutryn and terbuthylazine were significantly less reactive.

An ELISA for the triazine herbicide, cyanazine (*86b*) was recently reported in abstract form [300a)].

A detailed analysis of the specificity, sensitivity as well as the synthetic and immunological procedures for the immunoassay of the thiocarbamate herbicide, molinate (S-ethyl hexahydroazepine-1-carbothioate) (*86c*) is available [316a)] including a laboratory half-life study of rice field waters. The assay is specific for molinate with some (15%) cross reaction against molinate sulfone, but less than 1.4% cross reaction for structural relatives such as EPTC, cycloate, pebulate and vernolate [316a)].

An ELISA for chlorsulfuron was reported by Kelley et al. [173a)]. This assay was applied to the analysis of soil extracts for the herbicide. Preliminary data on a MAB against surflan (3,5-dinitro-N^4,N^4-dipropylsulfanilamide) (*86d*) were reported by Kuniyuki and McCarthy [180c)]. Their report is interesting, because a) the beaded affinity matrix, Affi-gel 10, an agarose based support containing a relatively long spacer carrying N-hydroxysuccinimide-activated carboxyl groups, was used as immunogenic carrier in Balb/c mice and b) the antigen, surflan (*86d*), was directly adsorbed, from a solution in DMSO, onto the surface of standard polystyrene microtitration plates. Both strategies might be useful in handling immunogen synthesis and antigen-coating for other antigens, too, which are only sparingly soluble in aqueous solution.

Representative assays discussed here are listed in Table 7 including assay formats and some characteristic assay parameters.

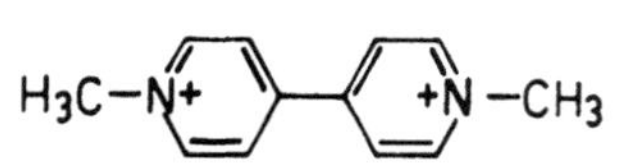

(83)

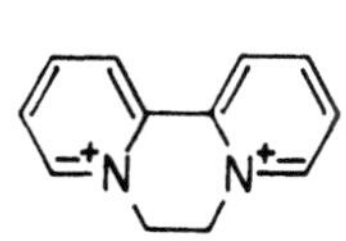

(84)

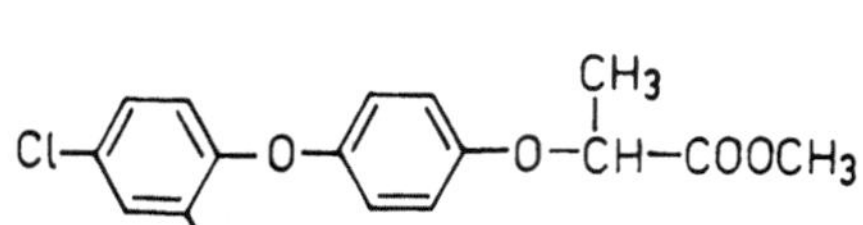

(85)

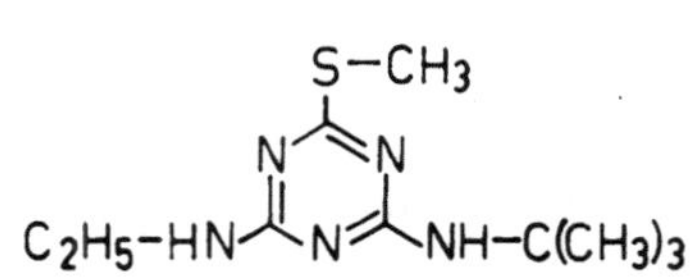

(86)

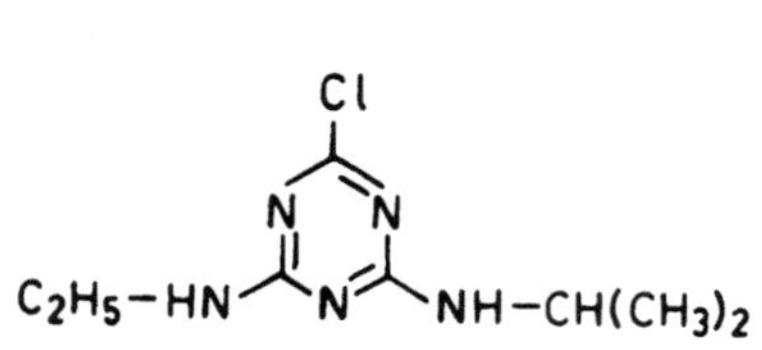

(86a)

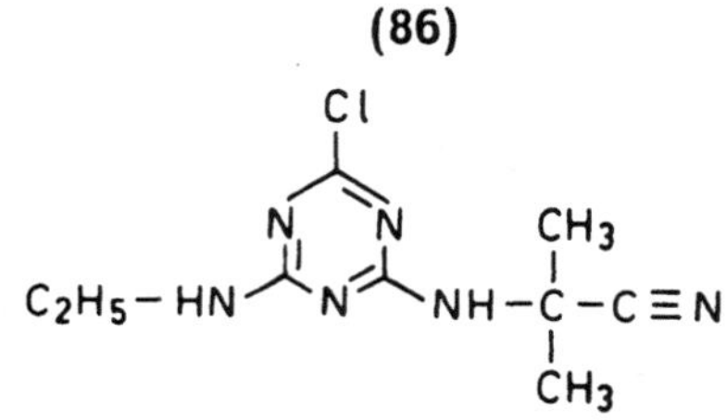

(86b)

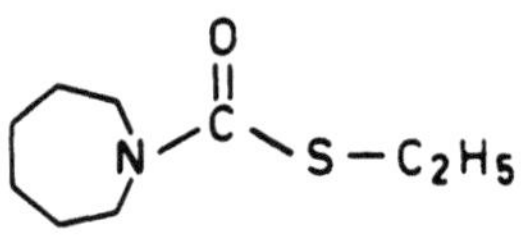

(86c)

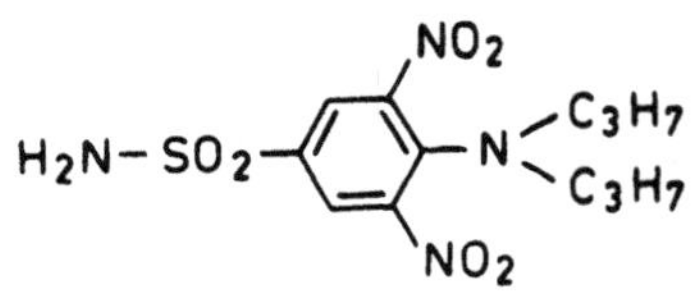

(86d)

7 Immunoassays in Food and Environmental Analysis

7.1 Scope

The environmental and food screening for unwanted or potentially hazardous constituents faces very much the same problems as were discussed in chapters 4 and 5, i.e. discrimination of a low level of agent against a high background of interference in very complex samples. Samples can be very diverse and range from airborne material from filters to soil samples, feces, human body fluids and clothes (environmental analysis) and from aqueous solutions to highly processed meat, flour or vegetable and fruit diets (food analysis).

Immunoassays in food analysis have already found considerable interest and a recent review on the history of this analytical technique in food science counted 160 publications (see chapter 1 of [243]). These include: the detection of bacteria and bacterial toxins, hormones including synthetic analogues, mycotoxins, parasites and viruses, in addition to proteins and other endogenous food components like caffeine, glycoalkaloids, bitter principles (see Chapt. 3.2.) as well as food additives or contaminants such as antibiotics, surfactants and pesticide residues. A bibliography and various reviews on the subject can be found in [243]. Addtional reviews cover this subject [65, 250] and especially the analysis of mycotoxins [45, 51, 55, 65] or toxins in general [55, 65]. Although RIA has been most frequently employed, the design of non-isotopic immunoassays in food analysis is of special importance for use in the food industry. The use of immunoassays is especially attractive in food analysis, because usually, large series of samples have to be screened for one or a small number of contaminat(s). A good example for this is the enzyme immunological study of the *Citrus* bitter principles, naringin (*40*) and limonin (*13*), in Florida test house juice samples [228]. ELISA allows an on-line check for juice bitterness level in fruit processing plants. The ease with even low levels of contaminations such as e.g. mycotoxins or anabolic hormones can now be detected should improve foodsafety considerably. The availability of an additional, independent technique to validate "positive" samples, is however, highly desirable because like any other technique, immunoassay may also give "false positives". This was shown for diethylstilbestrol (*95*) immunoanalysis which may be susceptible to interference by fluorinated corticosteroids [31].

For environmental analysis, immunoassay certainly will prove most useful, as well, although the application of this technique is as yet not widely recognized. Examples are, e.g. the analysis of spray drifts of certain pesticides and pesticide residues, exposure studies, the screening of surface water contamination by auxin herbicides (see Chapt. 4.2.), the analysis of some widespread carcinogens or mutagens, and also clinical studies. A certain limitation of the application of immunoassay in environmental testing may lie in the fact that it is often not known which contaminant(s) may be present. In this situation, physico-chemical analysis, such as mass-spectroscopy, is indispensable.

7.2 Mycotoxins

Most available assays were designed for the analysis of aflatoxins (7.2.1), trichothecenes 7.2.2) but also others (7.2.3). Table 8 gives a compilation of available techniques and lists assay formats and characteristic assay parameters.

Table 8. Selected immunoassays for the determination of mycotoxins. The following abbreviations are used: CL, complement-mediated lysis of liposomes; CF, carboxyfluorescein; all others as in Tables 4 and 5. If not stated otherwise, data are for standards

Compound	Assay format	Detection limit Measuring range	Reproducibility	Applications	Refs.
1. Aflatoxins					
AFB_1 (*87*)	E,P,SP,PO	M 1—100 ng/ml	W 22, B 21	Naturally contaminated corn	290
	E,P,SP,PO	M 1—20 ng/ml	W < 23 (corn)	Naturally contaminated corn	316 c
	E,P,SP,PO	M 0.05—1 ng/ml	B < 10 (corn, peanut)	Naturally contaminated corn and peanut butter	46 a
AFM_1	E,P,SP,PO	M 0.05—1 ng/ml	—	—	277 a
	(a) E,P,SP,PO	D 0.5 ng/ml (milk) M 0.05—10 ng/ml	—	Fortified and naturally contaminated milk	274
	(b) R,P,LP,T	D 0.25 ng/ml (milk) M 5—50 ng/ml	—		
	E,P,SP,PO	M 0.05—0.5 ng/ml D 5—10 ppt (milk) 25—50 ppt (cheese)	W 4.5—16 (milk)	Fortified milk and cheese	101
2. Trichothecenes					
T-2 (*88*)	E,P,SP,PO	M 0.1—1.0 ng/ml		—	277 a
	E,M,LP,CL,CF	D 2 ng/assay M 2—2000 ng/assay	—	—	206 b
	R,P,LP,T	M 0.1—5 ng/assay	—	Spiked corn	358 a
	R,P,LP,T	D 0.025—0.05 ng/assay M 0.05—2 ng/assay	—	—	402
Deoxynivalenol	R,P,LP,T	D 0.1 ng/assay M 0.1—2.5 ng/assay	—	Naturally contaminated grain samples	398 a

Table 8. (continued)

Compound	Assay format	Detection limit Measuring range	Repro-ducibility	Applications	Refs.
Diacetoxyscirpenol (*90*)	E,M,SP,PO	M 0.1—100 ng/ assay	—	—	261 a
	R,P,LP,T	D 0.06—0.4 ng/ assay M 5—500 ng/ assay	—	—	402
Roridin A (*90a*)	E,P,SP,PO	D 5 ng/ml M 5—25 ng/ml	W < 5	—	217 a
3. Other mycotoxins					
Zearalenone (*91*)	R,P,LP,T	D 5 ppb (human serum) M 0.25—10 ng/ assay	W 2—12	Fortified human serum	277 a
	E,M,SP,PO	D 0.5 ng/ml	W 3.3, B 4	Fortified corn	70 a
Ochratoxin A (*92*)	E,P,SP,PO	D 25—50 pg/ assay M 25—1250 pg/ assay	—	Wheat	196
	E,P,SP,AP	M 0.01—10 ng/ assay D 60 ng/kg (barley)	—	Fortified and naturally contaminated barley	239
Zinniol (*93a*)	R,P,LP,T	D 0.14 ng/ assay M 0.2—10 ng/ assay	W 10, B 14	—	235 a

7.2.1 Aflatoxins

The highly toxic and carcinogenic aflatoxins produced by *Aspergillus flavus* and *A. parasiticus* occur worldwide in corn, peanuts and other human foods as well as animal feedstuffs. Aflatoxin M_1, a metabolite of aflatoxin B_1 (*87*) may occur in milk and dairy products when animals are fed contaminated diet. Carcinogenic properties and acute toxicity of aflatoxins have been related to aflatoxin metabolites formed by hepatic microsomal enzymes. In this process, aflatoxin B_1 (*87*, AFB_1) binds to DNA and forms 2,3-dihydro-2-(N7-guanyl)-3-hydroxy-AFB_1 (AFB1-N^7-Gua) which converts, under physiological conditions, to the persistent adduct: 2,3-dihydro-2-(N^5-formyl-2′,5′,6′-triamino-4-oxo-N5-pyrimidyl)-3-hydroxy AFB_1. In addition, 2,3-dihydro-2,3-dihydroxy-AFB_1 is formed which may react, via Schiff's bases, with protein amino groups. The concern about aflatoxin-induced cancer has led to the development of an array of immunological techniques to detect and quantitate the free aflatoxins B_1 (*87*)[21,39,46a,49,81,90,116,190,192,270,272,277a,290,316c], B_{2a},[109,184,275], M[101,116,131,152,270,274,277a,397] as well as aflatoxin DNA adducts[115,116,135,139,271,275]. Recently monoclonal antibodies against aflatoxins[39,116,397] and their DNA-adducts[115,116,135,139] were also developed. Groopmann et al. used monoclonal antibody immunosorbent columns to enrich aflatoxins from biological sources[116] and Pestka and coworkers localized aflatoxin B_1 (*87*) in rat liver using the indirect immunoperoxidase technique[269].

Functionalization of the aflatoxin molecule has been achieved via the hemiglutarate (aflatoxin B_{2a}) [184] or, more commonly, via the 1-O-carboxymethyloxim of aflatoxin B_1 (*87*) or M_1 [39,48,90,290]. Aflatoxin B_1-diol (AFB_1-diol) was coupled, through Schiff's bases to ethylene-diamine-substituted BSA and the adduct was stabilized by borohydride reduction [271]. In the latter case, AFB_1-diol to BSA ratios of 6:1 were found optimal to elicit high serum titers in rabbits. Cross reactions were observed with aflatoxin B_1 (*87*), aflatoxin B_{2a} and aflatoxin M_1, whereas aflatoxin G_1, G_2, Q_1 and AFB_1-N^7-Gua were only slightly cross reactive. The antisera did recognize AFB_1 modified DNA but not unmodified DNA [271].

Monoclonal anti-AFB_1 antibodies [397] were found to be less specific than rabbit sera and recognized, relative to AFM_1, AFB_1 (*87*), AFG_1 and AFG_2, better than the sera did. Monoclonal antibodies raised against AFB_1 (*87*) also recognized AFG_1, AFB_2, AFM_1 and AFM_2, although about ten- to twenty fold more competitor was required relative to AFB_1 (*87*). In contrast, anti-AFB_1 monoclonal antibodies raised in a different study [116] recognized AFB_1 (*87*) and AFB_2 to the same extent. By further selection, it seems probable that considerably more selective MAB may still be found. On the other hand, for a first screening for potentially contaminated samples, a less specific antibody may initially be better suited. With ELISA as little as 0.1 ng ml^{-1} pure aflatoxin may still be detected [101,274] although sensitivity may be lower in complex samples.

The analysis of aflatoxins from biological samples usually involves enrichment and clean-up on reserved phase (C_{18}) cartridges which bind the toxins. These are then eluted with aqueous methanol or acetonitrile. This phase, after removal of the solvents and addition of assay buffer, may then be analyzed [101]. Milk may be analyzed for AFM_1 directly [274]. A recently described ELISA allows to quantitate AFB_1 in naturally contaminated agricultural commodities such as cornmeal or peanut butter. Quantitative results in the ppb-range are obtained within one hour (assay time). The samples are extracted with 70% methanol and 1% dimethylformanide in water, diluted with assay buffer to a final ethanol concentration of 7% and analyzed as such. The reference TLC assay correlated with the ELISA technique (slope 0.99, $r = 0.91$) when naturally contaminated corn samples were analyzed [46 a)].

A further step towards simplifying assay protocols will come from dipstick technologies. Here, the antibody, immobilized on a suitable support is supplied, in controlled amount, on a mechanical carrier (dipstick). This device is incubated in the sample of unknown antigen concentration. Residual, unoccupied antigen binding sites can be titrated after this incubation step by a variety of techniques such as, e.g. incubation with enzyme linked antigen, or fluorescent antigen. As a step towards this goal, Singh and Jang [316 c)] reported on a membrane-based enzyme immunoassay for AFB_1 whose sensitivity could be adjusted within a certain limit to a desired level by manipulation of the amount of antibody immobilized on the membrane (Immobilon™) [316 c)].

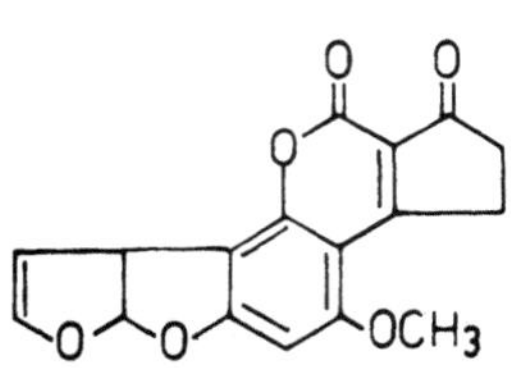

(87)

7.2.2 Trichothecenes

Trichothecenes are toxic fungal secondary metabolites produced by members of genera *Fusarium*, *Trichoderma*, *Myrothecium* and *Stachybotrys* which are frequently found as contaminants of foods and feeds. The trichothecene toxin, T-2 (*88*), is a potent protein biosynthesis inhibitor and is the most toxic trichothecene thus far isolated. In addition to T-2 toxin (*88*), deoxyverrucarol (*89*, DOVE) and diacetoxyscirpenol (*90*, DAS) and deoxinivalenol (DON) which are also suspected toxins, may occur. T-2 toxin (*88*) causes the highly fatal "alimentary toxic aleukia" and thus, aside from food monitoring, clinical applications notably expoiting the toxin-neutralizing effect of trichothecene antisera [95)] exist for anti-trichothecene antibodies. Both, RIA [99,139 a, 197,402)] and ELISA [91,91 a,110,273,277 a,278)] techniques were developed for the T-2 toxin (*88*) [47,91,95,99,110,154,195,197,273,277 a,278,402)] and for the related compounds, DOVE (*89*) [50)], DAS (*90*) [46,261 a,402)] and DON [398 a)] with detection limits in the range of 25—60 pg [402)]. In order to render trichothecenes immunogenic, the hemisuccinate of T-2 toxin and the hemiglutarate of DAS were employed as haptens but, perhaps due to instabilities in the animal, only low titer sera were elicited. Zhang et al. [402)] have synthesized the 3-carboxymethyloximes of both T-2 toxin and DAS and linked them to BSA. These conjugates elicited high titer sera which permitted the detection of 50–60 pg of the respective trichothecenes, by RIA, although there was cross reaction by some T-2 metabolites. Antibodies raised against DAS showed a high degree of cross-reactivity with monoacetoxyscirpenols. A significant improvement was recently reported by Pauly et al. [261 a)] who have succeeded in raising monoclonal antibodies to DAS. Immunization was done with DAS-hemiglutarate-BSA conjugates. High titers of anti-DAS MAB were obtained in ascites fluid. More important, the specificity of their MAB was higher than that reported for antisera so far. In addition to DAS (*90*), triacetoxyscirpenol and 15-monoacetoxyscirpenol were highly reactive. A significant cross-reaction against diacetylverrucarol was also observed but, among others, 4-monoacetoxyscirpenol, verrucarol, deoxynivalenol, T-2 toxin and trichothecin were only slightly or not reactive. The detection limit of the assay for standard DAS (*90*) was 1 ng. It is hoped that further work as to the application of this assay to the analysis of biological specimens will be available soon.

Assays have been applied to the analysis of T-2 (*88*) in toxin producing *Fusarium* ssp. [245 c)], in *Fusarium*-infected corn [110,195)] and wheat [195)] and of DON in wheat and corn [398 a)]. Group A-selective trichothecene antibodies were also reported recently [359)]. Rabbit antisera conjugates of hemisuccinate derivative of the macrocyclic trichothecene, roridin A (*90a*), a cytostatic compound produced by *Myrothecium verrucaria* and *M. roridum* which is held responsible for livestock poisoning, through contaminated feed, in Brasil and Argentinia [217 a and refs. herein)], were used to develop an ELISA for the 5–25 ng/ml concentration range [217 a)]. Cross-reactions were reported to occur with roridin J and, to a lesser extent, with verrucarin A and satratoxin H and G. Applications to biological samples were not reported.

7.2.3 Other Mycotoxins

Zearalenone (*91*) is an estrogenic mycotoxin which may occur in cereals and it causes hyperestrogenism in animals, e.g. swine, fed on contaminated diet. Liu et al. [208)] have synthesized the 6′-carboxymethyl-oxime of zearalenone and coupled this to BSA.

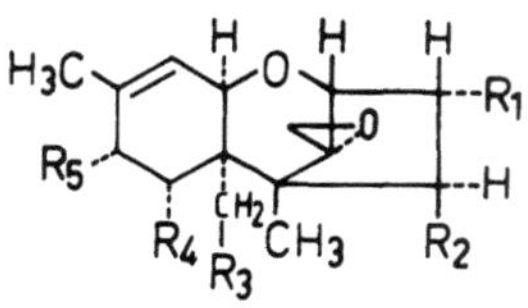

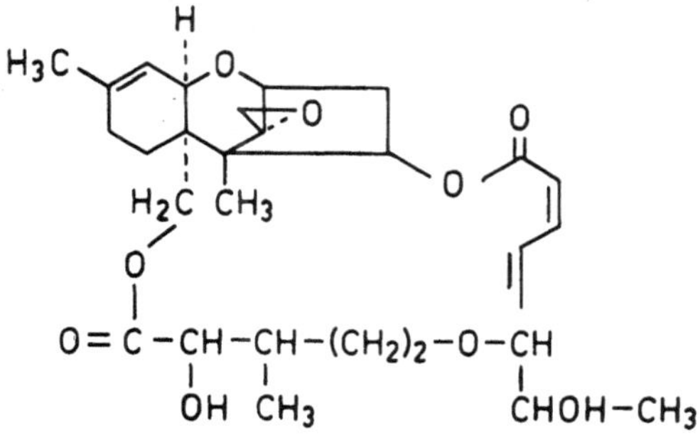

(90a)

	R_1	R_2	R_3	R_4	R_5
(88)	OH	OAc	OAc	H	$OCOCH_2CH(CH_3)_2$
(89)	H	H	OH	H	H
(90)	OH	OAc	OAc	H	H

High antibody titers were obtained in rabbits [208] but swine also responded [271, 336 a], a fact which may be exploited in the prophylaxis against zearalenone-caused hyperestrogenism in this species.

By competitive indirect ELISA, Liu et al. [208] were able to analyse the toxin in the range from 1—50 µg/ml. Some cross-reaction occurred against α-zearalenol (50%), β-zearalenol (12%) and α- and β-zearalenol (<10%). Zearalenone (*91*) was detected in aqueous methanolic samples of corn, wheat and pig feed [208]. Very recently, MAB specific for zearalenone were described and this might prove to be useful in future studies [70, 70 a].

The dihydro-coumarin derivative, ochratoxin A (*92*), a secondary metabolite of several fungal species including *Aspergillus* and *Penicillium*, probably causes mycotoxicoses in animals and possibly in humans. Ochratoxin A (*92*) has been linked closely with e.g. mycotoxic porcine nephropathy and maybe involved in the often fatal "Endemic Balkan Nephropathy" found in parts of Bulgaria, Roumania and Yugoslavia (see Refs. in [240]). Thus, in order to help ochratoxin A (*92*) detection in agricultural commodities as well as for clinical monitoring of ochratoxin, antisera have been produced and RIA as well as direct and indirect ELISA techniques have been developed. Aalund et al. coupled the toxin, using its free carboxyl group, to bovine IgG with EDC and immunized rabbits with this conjugate[1]. This RIA detects ng amounts of the toxin although at a low precision. ELISA techniques were found to be much more sensitive and allow permit to detection of between 25 pg [277] and 10 pg/assay [238] of ochratoxin A (*92*). Unfortunately, applicability in biological samples was not shown in these studies. Later, Lee and Chu [196] applied ochratoxin A ELISA to the analysis of wheat samples. At toxin levels below 30 ppb, cleanup of samples was necessary. This was conveniently done by reversed phase (C_{18}) cartridge chromatography of dilute aqueous methanol samples. This technique lowered the limits of detection to 1–2 ppb with recoveries in the range from 85–90% (variation coefficients 10–15%). An extremely specific ochratoxin A antiserum (cross reaction with ochratoxin B (*93*): 0.5%, ochratoxin α: 2.4%, coumarin: <0.00005%) was reported by Morgan et al. [240]. Ochratoxin A (*92*) could be detected in crude barley extracts and the sensitivity was 60 pg/g barley [239,240]. The phytotoxic *Alternaria* toxin, zinniol (3-methoxy-4-methyl-5[3-methyl-2,2-butenyloxy]-1,2-benzene dimethanol (*93a*), was measured by RIA [235 a)]. Monosuccinyl-zinniol was coupled to BSA and the conjugate used to immunize rabbits. The RIA detected from 0.05–50 µg of the toxin per assay and proved highly selective [235 a)].

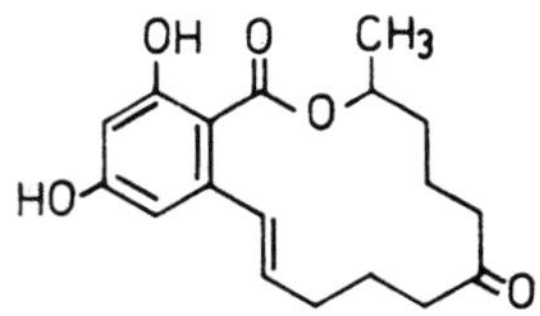

(91)

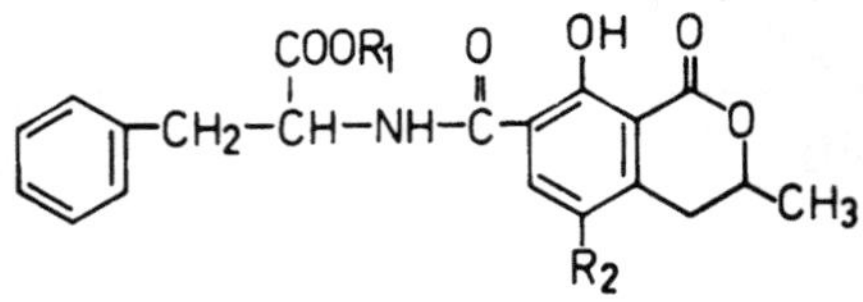

(92) $R_1 = H$ $R_2 = Cl$

(93) $R_1 = H$ $R_2 = H$

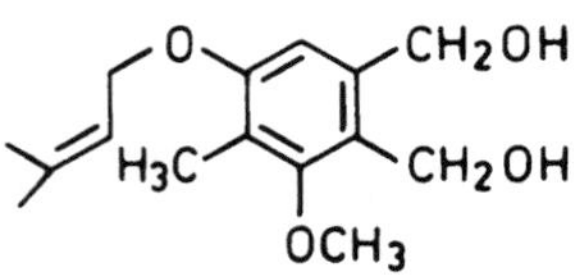

(93a)

7.3 Synthetic Chemicals

Environmental or food analysis for synthetic chemicals includes mutagenic or carcinogenic compounds, poisons and anabolic hormones.
The organophosphorous cholinesterase inhibitor, soman (*94*), (methylphosphonofluoridic acid 1,2,2-trimethylpropyl ester), a powerful nerve agent, was coupled to BSA and KLH by diazotization using a *p*-aminophenyl spacer [155]. The link was through the phosphono group and results in a stable, immunogenic conjugate. Monoclonal antibodies were selected by ELISA and this assay detected as little as 200 ppb free soman (*94*). Lenz et al. [201] compared the specificities of monoclonal and polyclonal anti-soman antibodies. The authors found MAB to be more selective than the polyclonal sera. Although an ELISA has been reported, its sensitivity was rather low (inhibition required > 10 µM soman) due to the low affinity of the MAB. However, this MAB was designed primarily to protect acetylcholinesterase from soman attack and was shown to be a potential prophylactic agent against soman toxicity [201].

The synthetic anabolic hormone, diethylstilbestrol (*95*, DES), has been banned throughout the European Community because it is a suspected carcinogen. To detect illegal application, the monitoring of slaughterhouse output and suspect meat products by immunological assays has become very attractive. Anti-DES sera were produced by several groups and some are commercially available [107,162,358] and are highly selective for DES (*95*) (see [358]). Sera are used in RIAs and allow DES determinations at the pg/g level [108,145,348,358]. Most critical has been the rather high and variable background signal obtained from untreated samples (see, e.g. [358]). This has led to various attempts to devise simple, yet efficient, pre-assay protocols such as

Table 9. Selected immunoassays for synthetic chemicals of environmental concern. For abbreviations, see the legends of Tables 4 and 5. If not stated otherwise, data are for standards. a = estimated, actual values not given

Compound	Assay format	Detection limit Measuring range	Repro-ducibility	Applications	Refs.
Soman (*94*)	E,M,SP,AP	D 200 ppb (1 × 10^{-6} M) M 10^{-3} – 10^{-6} M	—	—	155
Diethylstilbestrol (*95*)	R,P,LP,T	D 0.07 ng/g (meat)	—	fortified meat	348
	R,P,LP,T	M 10—500 pg/assay[a]	W 3—42 (urine)	fortified urine	108
2,3,7,8-Tetra-chlorodibenzofuran (*97*)	R,P,LP,I	D 30 pg/assay M≈ 50 pg—20 ng[a]	—	—	5
	R,P,LP,I	M 0.02—4 ng/assay	—	animal toxi-city study	214
Benzo[α]pyrene (*98*)	R,P,LP,T	M 0.1—4 ng/assay D 0.1 ng/assay	W 1.2—6.5	fortified human serum	168
4-Acetamido-biphenyl (*99*)	R,P,LP,I	M 0.3—190 pmol/assay D 0.03 pmol/assay	W 5—7 B 7—14	clinical	161

Sephadex LH-20 chromatography [348] or cartridge (C_{18}) chromatography [108, 348]. False-positives were also reported for DES RIA in samples from animals after the application of fluorinated corticosteroids [31]. The immunoassay of anabolic hormones in meat was recently reviewed [358]. Another application of immunoassay is the quantitation of the synthetic anabolic compounds, zeranol (*96*) [71, 72]. This compounds bears structural similarities to the mycotoxin zearalenone (*91*) which may cross react significantly in the zeranol RIA [71] based on sera obtained by immunizing sheep with zeranol-7-hemisuccinoyl-BSA conjugates.

For obvious reasons, environmental screening for mutagenic or carcinogenic chemicals, has become more important during recent years and immunological assays have been devised and tested for their suitability to monitor such compounds. Examples of this are 2,3,7,8-tetrachlorodibenzofuran (*97*), which may be a contaminant of commercial preparations of polychlorinated biphenyls, pentachlorophenol, or phenoxyacetic acid type herbicides [5, 214]. Benzo(α]pyrene (*98*), a polycyclic aromatic hydrocarbon found in coke oven and gas retorte environments [168], 4-acetamidobiphenyl (*99*), a metabolite of 4-aminobiphenyl which may occur as a contaminant of the raw chemical, diphenylamine [161], and 2-acetylaminofluorene, a model compound for investigating the role of DNA modifications by carcinogens [282].

2,3,7,8-tetrachlorodibenzofuran (*97*) antisera have been raised using 1-amino-3,7,8-trichlorodibenzo-*p*-dioxin which was linked, via an adipamide bridge, to thyroglobulin or albumins using the mixed anhydride technique [5]. About 30 pg of the antigen can be detected by RIA and there is some cross-reaction by dioxin isomers and other chlorinated dibenzofurans. A detailed specificity and assay validation study is reported in [214]. The same immunogen was administered to Balb/C mice and their spleen cells fused with SP2/0 myelma cells to produce hybridomas [344 b]. Four clones were selected which secrete dioxin-specific MAB useful in ELISA. These MAB retain their antigen-binding ability even in 25% DMSO. This feature allows to handle dioxin-containing samples and standard solutions conveniently and thus to minimize work-up-losses of this lipophilic antigen. Detailed specificity analyses for anti-dioxin MAB are available [344a)] which exemplify the promises and problems of using antibodies for the detection of trace chemicals.

Benzo[α]pyrene-6-isocyanate was conjugated to amino groups of BSA via a carbamido linkage [168)] to elicit antisera in rabbits. Most sera were of rather low titer but one serum (B4—3) gave a large response. This was characterized further and proved rather specific. Cross reactions in percent were: 3-OH benzo[α]pyrene (20), benz[α]-anthracene (10.5), pyrene (2), anthracene (1.4), phenanthrene (0.8). The assay range for benzo[α]pyrene (*98*) in RIA extended from 0.4–20 pmoles antigen/assay.

Antiserum against 4-acetamindobiphenyl (*99*) was obtained via 4-hemisuccinamidobiphenyl linked to thyroglobulin via the EDC technique. The serum proved very specific and recognized, in addition to 4-acetamidobiphenyl and N-hydroxy-4-acetamidobiphenyl, only 4-aminobiphenyl ($<3\%$). Other aminobiphenyls (e.g. the 2- and 3-isomers) were very slightly (0.1 %) reactive. The detection limit of the assay was given as ca. 6 pg (0.03 pmol) of 4-acetamidobiphenyl standard [161)]. Table 9 lists representative assays described in this chapter according to assay format and provides the characteristic assay parameters.

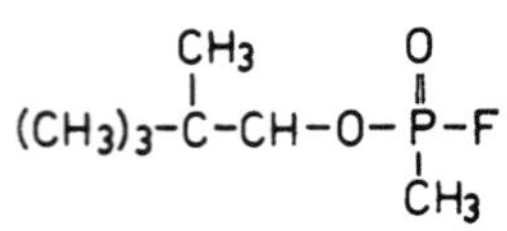

(94)

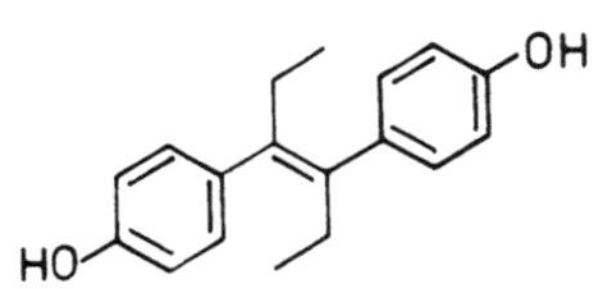

(95)

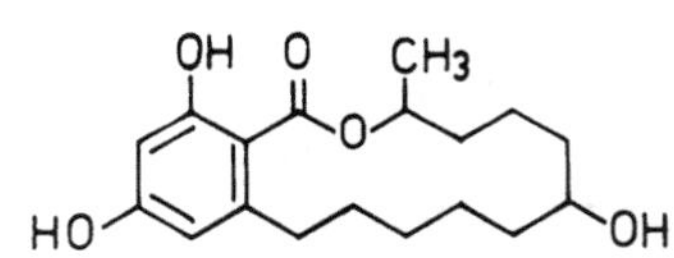

(96)

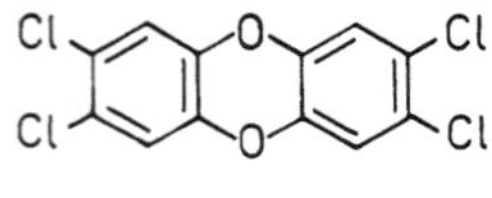

(97)

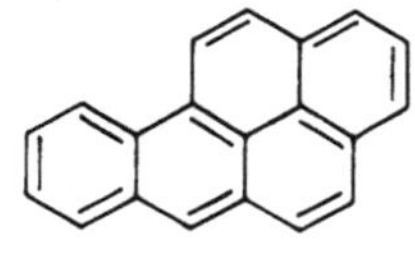

(98)

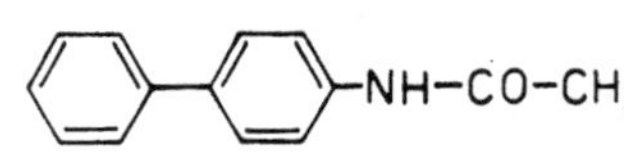

(99)

8 Conclusions

The immunological analysis of secondary plant products, plant growth regulators, pesticides and food or environmental contaminants is, albeit slowly, gaining ground. Immunoassays offer high sensitivity and often good selectivity. Their major advantages lie in solving screening problems or providing fast results when the compound of interest is known. This is frequently the case, especially in pesticide, food or environmental analysis. Immunoassays are of limited use in structural identification but may help to screen for classes of chemicals suspected to be present in a sample. Their major disadvantage is that assay development is a multi-step and somewhat lengthy procedure requiring immunological expertise. The development of non-isotopic immunoassays, based on monoclonal antibodies is the method of choice for the future. Hybridoma technology provides unlimited amounts of highly pure antibodies of predifined characteristics. Although hybridoma selection is a statistical process, proper design of the screening assays used at this stage, ensures that antibodies with just the right properties can be found. With the specific target of pesticide analysis in mind, perhaps the greatest benefit of using immunoassays is that such assays permit to monitor, in great detail, pesticide uptake and translocation in treated organisms, in the environment, in worker exposure studies and in toxicological studies in general. Immunological assays may furthermore prove very useful in helping the farmer optimize pesticide application programs, screen for residues in treated agricultural crops etc. With the advent and further development of "dip-stick-technology" [207,403] immunoassays may be carried out in the field and eventually become as simple as the "test-strip" analysis of many metabolites and pH-values already is. There is also an increasing engagement of industry to develop immunoassay kits for plant growth regulators, pesticides and food and environmental contaminants. A collection of assays using advanced technologies is already on the market. This area was recently reviewed and the interested reader is referred to reference [177a], for a survey.

Acknowledgements: My thanks are due to Prof. R. L. Mansell, Tampa, FL, for checking the English, to Göran Key, Osnabrück, and Hubert Winking, Bochum, for the artwork and to Dr. R. Stöcker, Monheim, FRG and Dr. R. Mertens, Berlin, FRG, for part of the literature search. This work was funded by the Fonds der Chemischen Industrie, Frankfurt, FRG (literature provision). The author's work cited herein was funded by grants of the Deutsche Forschungsgemeinschaft, Bonn, FRG, the Stiftung Volkswagenwerk, Wolfsburg, FRG and the Stifterverband für die Deutsche Wissenschaft, Essen, FRG.

9 List of Abbreviations

ABA	abscisic acid
BSA	bovine serum albumin
DAS	diacetoxyscirpenol
DES	diethylstilbestrol
DHZ	dihydrozeatin
DHZR	dihydrozeatin riboside
DON	deoxinivalenol
DOVE	deoxyverrrucarol
Dx	digitoxose
EDC	1-ethyl-3(3-dimethylaminopropyl)-carbodiimide HCl
EIA	enzyme immunoassay
ELISA	enzyme linked immunosorbent assay
EMIT	enzyme multiplied immunoassay technique
FIA	fluorescence immunoassay
GA_x	gibberellin A_x
Gal	galactose
GC	gas chromatography
GC-MS	gas chromatography-mass spectrometry

Glc	glucose
Glcr	glucuronic acid
HPLC	high pressure liquid chromatography
HSA	human serum albumin
IAA	indole-3-acetic acid;
IP	isopentenyladenine
IPA	isopentenyladenosine
KLH	keyhole limpet hemocyanin
LMC	low molecular weight compound
LSC	liquid scintillation counting
MAB	monoclonal antibody
MBC	methyl 2-benzimidazole carbamate
Neo	neohesperidosyl
OVA	ovalbumin
Rham	rhamnose
RIA	radioimmunoassay
Rib	ribosyl
THY	thyroglobulin
XA	xanthoxin
Z	zeatin
ZR	zeatin riboside

Notes added in proof

During processing of this manuscript, several papers appeared or came to our attention which have immediate relevance for this topic. These are listed below together with a brief description of their contents arranged according to the sequence of chapters in this article.

5 Growth Regulators

5.2 Auxins. Marcussen J et al. (1989) Plant Physiol. 89: 1071 (novel conjugation method for IAA via C5); Pelese F et al. (1989) Plant Physiol. 89: 86 (determination of several phytohormones in a naphthaleneacetic acid tolerant tobacco mutant); Pilate G, Sossountzov L, Miginiac E (1989) Plant Physiol. 90: 907 (determination of several phytohormones in the aquatic fern, Marsilea drummondii.

5.3 Cytokinins. Parker CW, Badenoch-Jones J, Letham DS (1989) J Plant Growth Regul. 8: 93 (new RIA for quantifying cis-zeatin and cis-zeatin riboside, and application of the assay to xylem sap samples); von Schwartzenberg K, Lutze K, Hahn H (1988) J Plant Physiol. 133: 529 (determination of several cytokinins in spruce needles by ELISA).

5.4 Gibberellins. Durley RC et al. (1989) Plant Physiol. 90: 445 (purification of several gibberellins by immunoaffinity chromatography); Smith VA, MacMillan J (1989) Plant Physiol. 90: 1148 (gibberellin purification and quantitation by immunoaffinity chromatography and RIA in conjunction with GC-MS).

5.5 Growth Inhibitors. Vernieri P et al. (1989) J Plant Physiol. 134: 441 (new monoclonal antibody against cis, trans (S) ABA).

6 Pesticides and Metabolites

6.4 Herbicides. Nagao M et al. (1989) J. Immunoassay 10: 1 (detailed characterization of three monoclonal antibodies against paraquat); Quing Xiao Li et al. (1989) Anal. Chem. 61: 819 (comparison of ELISA and GC techniques for molinate residue analysis).

7 Food and Environmental Analysis

7.2 Mycotoxins. Jelinek CF, Pohland AE, Wood GE (1989) J. Assoc. Off. Anal. Chem 72: 223 (review on the worldwide occurrence of mycotoxins in foods and fees); Pestka JJ (1988) J. Assoc. Off. Anal. Chem 71: 1075 (review on the surveillance of foodborne mycotoxins by immunochemical assays); Morgan MRA (1989) Tetrahedron 45: 2237 (review on ELISA assays for mycotoxin analysis).

7.2.1 Aflatoxins. Märtlbauer E, Terplan G (1985) Arch. Lebensmittelhyg. 36: 49 (ELISA with high sensitivity for aflatoxin M_1 in milk and milk powder).

7.2.2 Trichothecenes. Hack R, Märtlbauer E, Terplan G (1987) J. Vet. Med. B34: 538 (monoclonal antibodies against T-2 toxin); Klaffer U, Märtlbauer E, Terplan G (1988) Int. J. Food Microbiol. 6: 9 (antiserum based ELISA for the detection of diacetoxyscirpenol); Hack R, Klaffer U, Terplan G (1989) Lett. Appl. Microbiol. 8: 71 (production and characterization of a monoclonal antibody against diacetoxyscirpenol); Hack R, Märtlbauer E, Terplan G (1988) Appl. Environm. Microbiol. 54: 2328 (production and characterization of monoclonal antibodies against roridin A); Fan TSL et al. (1988) Appl. Environm. Microbiol. 54: 2959 (production and characterization of a monoclonal antibody cross-reacting with most group A trichothecenes, e.g. T-2 toxin and diacetoxyscirpenol).

7.2.3 Other Mycotoxins. Gareis M et al. (1988) Z. Lebensm. Unters. Forsch. 186: 114 (determination of ochratoxin A in mother's milk by HPLC, MS and ELISA); Märtlbauer E, Terplan G (1988) Arch. Lebensmittelhyg. 39: 133 (ELISA for the determination of ochratoxin A in swine serum).

10 References

1. Aalund O, Brunfeldt B, Hald B, Krogh P, Poulsen K (1975) Acta path. microbiol. scand. Sect. C 83: 390
2. Adamson SR et al. (1983) Biotechnol. Lett. 9: 573
3. Adler FL, Liu C-T (1971) J. Immunol. 106: 1684
4. Akiyoshi DE, Morris RO, Hinz R, Mischke BS, Kosuge T, Garfinkel DJ, Gordon MP, Nester EW (1983) Proc. Natl. Acad. Sci. USA 80: 407
5. Albro PW et al. (1982) Meth. Enzymol. 84: 619
6. Altshuler GL et al. (1986) Biotechnol. Bioengin. 28: 646
7. Andrieu JM, Manas S, Dray F (1975) In: Hilier SG, Griffiths K, Cameron EHD (Eds.) Steroid Immunoassay, Alpha Omega Publishing, Cardiff, p 189
8. Arens H, Stöckigt J, Weiler EW, Zenk MH (1978) Planta Medica 34: 37

9. Arens H, Zenk MH (1980) Planta Medica 38: 214
10. Arens H, Zenk MH (1980) Planta Medica 39: 336
10a. Arens H, Deus-Neumann B, Zenk MH: Planta Medica 1987: 179
11. Atzorn R, Weiler EW, Zenk MH (1981) Planta Medica 41: 1
12. Atzorn R, Weiler EW (1983) Planta 159: 1
13. Atzorn R, Weiler EW (1983) Planta 159: 289
14. Atzorn R, Weiler EW (1983) Planta 159: 7
15. Badenoch-Jones J, Letham DS, Parker CW, Rolfe BG (1984) Plant Physiol. 75: 1117
15a. Badenoch-Jones J, Parker CW, Letham DS (1987) J. Plant Growth Regul. 6: 159
16. Barthe GA, Stewart I (1985) J. Agric. Food Chem. 33: 293
16a. Barthe GA, Jourdan PS, McIntosh CA, Mansell RL (1987) J. Plant Physiol. 127: 55
16b. Barthe GA, Jourdan PS, McIntosh C, Mansell RL (1988) Phytochemistry 27: 249
17. Bartos F, Bartos D, Dolney AM, Grettie DP, Campbell RA (1978) Res. Commun. Chem. Pathol. Pharmacol. 19: 295
18. Beale MH, Hooley R, MacMillan J (1986) In: Bopp M (ed) Plant Growth Substances 1985. Springer, Berlin Heidelberg New York p 65
19. Berson SA, Yalow RS (1959) Ann. N.Y. Acad. Sci. 82: 338
20. Berson SA, Yalow RS (1968) Clin. Chim. Acta 22: 51
21. Biermann A, Terplan G (1982) Archiv f. Lebensmittelhyg. 33: 17
22. Bjercke RJ, Cook G, Rychlik N, Gijka HB, Van Vunakis H, Langone JJ (1986) J. Immunol. Meth. 90: 203
22a. Bjercke RJ, Cook G, Langone JJ (1987) J. Immunol. Meth. 96: 239
23. Bopp M, Weiler EW (1985) Naturwissenschaften 72: 434
24. Boudene C, Duprey F, Bohuon C (1975) Biochem. J. 151: 413
25. Bower JP, Cutting JGM, Van Lehyveld LJ (1984) In: Die rol van hormone in die bevordering von plant produksie. Proc. Biochem. Physiol. Workshop, Pretoria, p 7
26. Brandon DL, Corse JW, Layton LL (1979) Plant Physiol. 63(S): 82
26a. Brandon DL, Corse JW (1987) In: Fox JE, Jacobs M (eds) Molecular biology of plant growth control Alan R. Liss, New York, p 209
27. Brenner ML (1981) Annu. Rev. Plant Physiol. 32: 511
28. Brimfield AA, Lenz DE, Graham C, Hunter KW jr. (1985) J. Agric. Food Chem. 33: 1237
29. Brisson L, Vacha WEK, Ibrahim RK (1986) Plant Science 44: 175
30. Brodelius P, Deus B, Mosbach K, Zenk MH (1979) FEBS Lett. 103: 93
30a. Brooker G, Fleming JW, Ross EM (1982) Biochem. Biophys. Acta 685: 39
31. Brunn H, Fister P, Eigenbrodt E, Stojanowic V (1985) Fd. Chem. Toxicol. 23: 931
32. Buchman D, Miller D, Koch G (1985) Hybridoma 4: 173
33. Butler JE (1980) In: Maggio ET (ed) Enzyme-immunoassay CRC Press, Boca Raton, p 5
34. Butler VP, Chen JP (1968) Proc. Natl. Acad. Sci. USA 57: 71
35. Butler VP (1975) J. Immunol. Meth. 7: 1
36. Cahill DM, Weste GM, Grant BR (1986) Plant Physiol. 81: 1103
37. Cais M et al. (1977) Nature 270: 534
38. Campbell AM (1984) Monoclonal antibody technology, Elsevier, Amsterdam
39. Candlish AAG, Stimson WH, Smith JE (1985) Lett. Appl. Microbiol. 1: 57
40. Carlsson M et al. (1985) J. Immun. Meth. 79: 89
40a. Carter BA, Oliver DG, Jang L (1985) Food Technol. 39: 82
41. Castonguay A, Van Vunakis H (1979) Analyt. Biochem. 95: 387
42. Castro A, Grettie DP, Bartos F, Bartos D (1973) Res. Commun. Chem. Path. Pharmac. 6: 879
43. Cekan Z (1975) J. Steroid Biochem. 6: 271
44. Centeno ER, Johnson WJ, Sehon AH (1970) Int. Arch. Allergy Appl. Immunol. 37: 1
44a. Chevrier D, Guesdon J-L, Mazi J-C, Avrameas S (1986) J. Immunol. Meth. 94: 119
45. Chu FS (1984) J. Food Prot. 47: 562
46. Chu FS, Chen Liang MY, Zhang GS (1984) Appl. Environm. Microbiol. 48: 777
46a. Chu FS, Fan TSL, Zhang GS, Xu YC (1987) J. Assoc. Off. Anal. Chem. 70: 854
47. Chu FS, Grossman S, Wei RS, Mirocha CJ (1979) Appl. Environ Microbiol. 37: 104
48. Chu FS, Hsia MTS, Sun PS (1977) J. Assn. Offic. Anal. Chem. 60: 791
49. Chu FS, Ueno I (1977) Appl. Environ. Microbiol. 33: 125
50. Chu FS, Zhang GS, Williams MD, Jarvis BB (1984) Appl. Environm. Microbiol. 48: 781

51. Chu FS (1984) In: Kurata H, Ueno YJ (eds) Toxigenic fungi — their toxins and health hazard Elsevier, Amsterdam, p 234
51a. Cohen JD et al. (1987) Plant Physiol. 84: 982
51b. Colbert DL, Gallacher G, Ayling P, Turner GJ (1988) Clin. Chim. Acta 171: 37
52. Collignon A, Edelman L, Scherrmann JM (1984) Int. J. Nucl. Med. Biol. 11: 97
53. Constantinidou HA, Steele JA, Kozlowski TT, Upper CD (1978) Plant Physiol. 62: 968
54. Cook CE, Turine ME, Myers M, Amerson E, Kepler JA, Taylor GF (1976) Commun. Chem. Path. Pharm. 13: 497
54a. Coxon RE, Rae C, Gallacher G, Landon J (1988) Clin. Chim. Acta 175: 297
55. Crowther JS, Holbrook R (1980) Soc. Appl. Bacteriol. Symp. Ser. 8: 339
56. Crozier A (ed) (1983) The biochemistry and physiology of gibberellins, Praeger, New York, vol 1, 2
57. Crozier A, Loferski K, Zaerr JB, Morris RO (1980) Planta 150: 366
58. Crozier A, Sandberg G, Monteiro AM, Sundberg B (1986) In: Bopp M (ed) Plant Growth Substances 1985. Springer, Berlin Heidelberg New York, p 13
59. Cutting JG, Lishman AW, Hofman PJ, Wolstenholme BN (1986) Acta Hort. 175: 285
60. Cutting JG, Hofman PJ, Lishman AW, Wolstenholme BN (1984) Crop Prod. 13: 38
61. Cutting JG, Lishman AW, Hofman PJ, Wolstenholme BN (1984) S. Afr. Avocado Grs. Assn. Yrbk. 7: 93
62. Cutting JG, Lishman AW, Van der Hoven A, Wolstenholme BN (1983) Crop Prod. 12: 113
63. Daie J, Wyse R (1982) Anal. Biochem. 119: 365
64. Dandliker WB et al. (1973) Immunochemistry 10: 219
65. Daussant J, Bureau D (1984) Dev. Food Anal. Tech. 3: 175
66. Davis GC et al. (1986) In: Bopp, M (ed) Plant Growth Substances 1985. Springer, Berlin Heidelberg New York, p 44
66a. Dean PDG, Johnson WS, Middle FA (1985) Affinity chromatography. IRL Oxford
67. De Greef W, Dekegel H, Hamers R (1980) Arch. Int. Physiol. Biochim. 88: B134
68. Delaage MA, Hirn MH, De Reggi ML (1982) Meth. Enzymol. 84: 350
68a. Deus-Neumann B, Stöckigt J, Zenk MH: Planta Medica 1987: 184
69. Deverill I, Reeves WG (1980) J. Immunol. Meth. 38: 191
70. Dixon DE, Ram BP, Harmon JT, Moorman MA, Williams SS, Hart LP (1986) Abstr. Annu. Meet. Am. Soc. Microbiol. 86: 271
70a. Dixon DE, Warner RL, Ram BP, Hart LP, Pestka JK (1987) J. Agric. Food Chem. 35: 122
71. Dixon SN (1980) J. Vet. Pharm. Ther. 3: 177
72. Duchatel JP, Maghuin-Rogister G (1985) Ann. Rech. Veterin. 16: 93
73. Dudley RA et al. (1985) Clin. Chem. 31: 1264
74. Duff RG (1985) Trends in Biotechnol. 3: 167
75. Durley RC, Morris RO (1983) Plant Physiol. 72(S): 114
76. Dwenger A (1984) J. Clin. Chem. Clin. Biochem. 22: 883
77. Eberle J, Arnscheidt A, Klix D, Weiler EW (1986) Plant Physiol. 81: 516
78. Eberle J, Wang TL, Cook S, Wells B, Weiler EW (1987) Planta 172: 289
79. Eberle J, Yamaguchi I, Nakagawa N, Takahashi N, Weiler EW (1986) FEBS Lett. 202: 27
80. Eberle J (1986) Thesis, Universität Osnabrück, p 205
81. El-Nakib O, Pestka JJ, Chu FS (1981) J. Assn. Offic. Anal. Chem. 64: 1077
82. Engvall E, Perlmann P (1971) Immunochemistry 8: 871
83. Ercegovich CD, Vallejo RP, Gettig RR, Woods L, Bogus ER, Mumma RO (1981) J. Agric. Food Chem. 29: 559
84. Erlanger BF, Beiser SM (1964) Proc. Natl. Acad. Sci. USA 52: 68
85. Erlanger BF (1980) Meth. Enzymol. 70: 85
86. Ernst D, Oesterhelt D (1985) Plant Cell Reports 4: 140
87. Ernst D, Schäfer E, Oesterhelt D (1983) Planta 159: 222
88. Ernst D, Schäfer W, Oesterhelt D (1983) Planta 159, 216
89. Ernst D (1986) In: Linskens HF, Jackson JF (eds), Immunology in Plant Sciences. Modern Methods of Plant Analysis. Springer, Heidelberg Berlin New York, vol 4 p 18
89a. Fackler U, Huber W, Hock B (1986) Forstw. Cbl. 105: 254
89b. Fackler U, Reich J, Hock B (1986) J. Plant Physiol. 126: 163
90. Fan TSL, Chu FS (1984) J. Food Prot. 47: 263

91. Fan TSL, Zhang GS, Chu FS (1984) J. Food Prot. 47: 964
91a. Fan TSL, Xu YC, Chu FS (1987) J. Assn. Offic. Anal. Chem. 70: 657
92. Faraj BA, Mu J-Y, Lewis MS, Wilson JP, Israili ZH, Dayton PG (1975) Proc. Soc. Exp. Biol. Med. 149: 664
93. Fatori D, Hunter WN (1980) Clin. Chim. Acta 100: 81
94. Feldmann H, Rodbard D (1971) In: Odell WD, Daughaday WH (eds), Principles of Competitive Protein Binding Assays. Philadelphia, p 57
94a. Ferguson BS (1986) Abstr. 5C-09, 6th Intern. Congr. Pest. Chem., IUPAC, Ottawa
95. Feuerstein G, Powell JA, Knower AT, Hunter KW jr. (1985) J. Clin. Invest. 76: 2134
95a. Feyerabend M, Weiler EW (1988) Physiol. Plant. 74: 181
95b. Feyerabend M, Weiler EW (1988) Physiol. Plant. 74: 185
96. Findlay JWA, Butz RF, Welch RM (1977) Res. Comm. Chem. Pathol. Pharmacol. 17: 595
97. Findlay JWA, Butz RF, Welch RM (1976) Life Sciences 19: 389
97a. Fliniaux MA, Jacquin-Dubreuil A: Planta Medica 1987: 87
98. Fong K-LL, Ho DHW, Carter CJK, Brown NS, Benjamin RS, Freireich EJ, Odey GP (1980) Anal. Biochem. 195: 281
99. Fontelo PA, Beheler J, Brunner DL, Chu FS (1983) Appl. Environm. Microbiol. 45: 640
100. Freeman LM, Blaufox MD (1975) Radioimmunoassay. Grune and Stratton, New York
101. Fremy JM, Chu FS (1984) J. Assn. Offic. Anal. Chem. 67: 1098
102. Fuchs S, Fuchs Y (1969) Biochim. Biophys. Acta 192: 528
103. Fuchs S, Haimovich J, Fuchs Y (1971) Eur. J. Biochem. 18: 384
104. Fuchs Y, Gertman E (1974) Plant Cell Physiol. 15: 629
105. Fuchs Y, Mayak S (1972) Planta 103: 117
105a. Fusseder A, Wagner B, Beck E (1988) Botanica Acta 101: 214
106. Garve R, Luckner M, Vogel E, Tewes A, Nover L (1980) Planta Medica 40: 92
107. Gaspar P, Duchatel J-P, Maghuin-Rogister G (1983) Ann. Md. Vt. 127: 451
108. Gaspar P, Maghuin-Rogister G (1985) J. Chromatogr. 328: 413
109. Gaur P, Lan HP, Pestka JJ, Chu FS (1981) Appl. Environm. Microbiol. 41: 478
110. Gendloff EH, Pestka JJ, Swanson SP, Hart LP (1984) Appl. Environm. Microbiol. 47: 1161
110a. Getz H-P, Schulte-Altedorneburg M, Willenbrink J (1987) Planta 171: 235
111. Gintzler AR, Mohacsi E, Spector S (1976) Eur. J. Pharmacol. 38: 149
111a. Glagasigij U, Sato Y, Suzuki Y (1988) Chem. Pharm. Bull. 36: 1086
112. Gochman N, Bowie LJ (1977) Analyt. Chem. 9: 1183A
113. Goding JW (1983) Monoclonal antibodies: principles and practice, Academic, London
114. Grisebach H, Moesta P, Hahn MG (1986) In: Linskens HF, Jackson JF (eds) Immunology in Plant Sciences. Modern Methods of Plant Analysis, New York, Springer, Berlin Heidelberg New York, vol 4 p 75
115. Groopman JD et a. (1982) Cancer Res. 42: 3120
116. Groopman JD et al. (1984) Proc. Natl. Acad. Sci. USA 81, 7728
117. Grossmann K, Schmidt HO, Jung J (1986) Plant Cell Reports 5: 315
118. Grossmann K, Weiler EW, Jung J (1985) Planta 164: 370
118a. Grossmann K, Kwiatkowski J, Siebecker H, Jung J (1987) Plant Physiol. 84: 1018
119. Grota LJ, Brown GM (1976) Endo 98: 615
120. Gushaw JB, Hu, MW, Singh P, Miller JG, Schneider RS (1977) Clin. Chem. 23: 1144
121. Haas GJ, Guardia EJ (1968) Proc. Soc. Exp. Biol. Med. 129: 546
122. Hagimori M, Matsumoto T, Kisaki T (1980) Plant Cell Physiol. 21: 1391
123. Hagimori M, Matsumoto T, Mikami Y (1984) Plant Cell Physiol. 25: 947
124. Hagimori M, Matsumoto T, Obi Y (1982) Plant Physiol. 69: 653
125. Hagimori M, Matsumoto T, Obi Y (1982) Plant Cell Physiol. 23: 1205
126. Hahn EF, Fishmann JH, Fishman J (1979) J. Pharm. Pharmacol. 31: 250
127. Haines CF, Mahajan DK, Miljkovic D, Miljkovic M, Vesell ES (1974) Clin. Pharm. Therap. 16: 1083
128. Hallberg M, Welander M (1985) 12 th Int. Conf. Plant Growth Substances, Book of Abstracts, Heidelberg, p 70
129. Hammock BD, Mumma RO (1980) ACS Symp. Ser. 136: 321
129a. Hammock BD, Gee SJ, Cheung PYK, Miyamoto T, Goodrow MH, Van Emon J, Seiber JN (1986) 6th Intern. Congr. Pest. Chem., IUPAC, Ottawa, p 309

130. Hansen CE, Wenzler H, Meins F jr. (1984) Plant Physiol. 75: 959
131. Harder WO, Chu FS (1979) Experientia 35: 1104
132. Harris CC et al. (1979) Proc. Natl. Acad. Sci. USA 76: 5336
133. Harris M, Outlaw W jr, Mertens R, Weiler EW (1988) Proc. Natl. Acad. Sci. 85: 2584
134. Harris MJ, Dugger WM (1986) Plant. Physiol. 82: 339
134a. Harris MJ, Dugger WM (1986) Plant Physiol. 82: 1164
134b. Harrison RO, Brimfield AA, Hunter Jr, Nelson JO (1986) Abstr. 5C-12, 6th Intern. Congr. Pest. Chem., IUPAC, Ottawa
134c. Hartung W, Radin JW, Hendrix DL (1988) Plant Physiol. 86: 908
134d. Hassan F., Rothnie NE, Yeung SP, Palmer MV (1988) J. Agric. Food Chem. 36: 398
135. Haugen A et al. (1981) Proc. Natl. Acad. Sci. USA 78: 4124
136. Heinemann WR, Anderson CW, Halsall HB (1979) Science 204: 865
137. Heinemann WR, Halsall HB (1985) Analyt. Chem. 57: 1321A
138. Heinemeyer W, Buchmann I, Tonge DW, Windass JD, Alt-Moerbe J, Weiler EW, Botz T, Schröder J (1987) Molec. Gen. Genet. 210: 156
138a. Heldman E, Balan A, Horowitz O, Ben-Zion S, Torten M (1985) FEBS Lett. 180: 243
139. Hertzog PJ, Lindsay Smith JR, Garner RC (1982) Carcinogenesis 3: 825
139a. Hewetson JF, Pace JG, Beheler JE (1987) J. Assn. Offic. Anal. Chem. 70: 654
140. Hirai N (1986) In: Takahashi N (ed) Chemistry of Plant Hormones, CRC, Boca Raton, p 201
141. Hirsch AM, Pengelly WL, Torrey JG (1982) Bot. Gaz. 143: 15
142. Hocart CH, Parker CW, Badenoch-Johnes J, Letham DS, Summons RE, Palni LMS (1985) 12th Int. Conf. Plant Growth Substances, Book of Abstracts, Heidelberg, p 68
142a. Hocart CH, Badenoch-Jones J, Parker CW, Letham DS, Summons RE (1988) J. Plant Growth Regul. 7: 179
143. Hodges CC, Rapoport H (1982) J. Nat. Proc. 45: 481
144. Hodgkinson AJ, Landon J, Smith DS, Sidki AM (1986) Ther. Drug Monitor. 8: 236
145. Hoffman B, Laschütza W (1980) Arch. Lebensmittelhyg. 31: 105
146. Hofman PJ, Featonby-Smith BC, Van Staden J (1986) J. Plant Physiol. 122: 455
147. Hofman PJ, Forsyth C, Van Staden J (1985) J. Plant Physiol. 121: 1
148. Horgan R (1981) Progr. Phytochem. 7: 137
149. Horgen PA, Nakagawa CH, Irvin RT (1984) Can. J. Biochem. Cell Biol. 62: 715
150. Hornberg C, Weiler EW (1984) Nature 310: 321
151. Hsu A-F, Brower D, Etskovitz RB, Chen PK, Bills DD (1983) Phytochemistry 22: 1665
152. Hu WJ, Woychik N, Chu FS (1984) J. Food Prot. 47: 126
152a. Huber SJ (1985) Chemosphere 14: 1795
153. Huber SJ, Hock B (1985) Z. Pflanzenkr. Pflanzensch. 92: 147
153a. Huber SJ, Sautter C (1986) J. Plant Dis. Protection 93: 608
153b. Huber SJ, Hock B (1985) GIT Fachz. Lab. 29: 969
153c. Huber SJ, Hock B (1986) In: Bergmeyer HU (ed) Methods of Enzymatic Analysis. VCH, Weinheim, vol 12 p 438
153d. Huhtikangas A, Lehtola T, Lapinjoki S, Lounasmaa M: Planta Medica 1987: 85
154. Hunter KW jr., Brimfield AA, Miller M, Finkelman FD, Chu FS (1985) Appl. Environm. Microbiol. 49: 168
155. Hunter KW jr., Lenz DE, Brimfield AA, Naylor JA (1982) FEBS Lett. 149: 147
156. Hunter KW jr., Lenz DE (1982) Life Sciences 30: 355
157. Hunter MM, Margolies MN, Ju A, Haber E (1982) J. Immunol. 129: 1165
158. Hurn BAL, Chantler SM (1980) Meth. Enzymol. 70: 104
159. Ishikawa E, Kato K (1978) Scand. J. Immunol. 8(S): 43
159a. Jaworski EG (1986) 6th Intern. Congr. Pest. Chem., IUPAC, Ottawa, p 585
160. Johannsson A, Stanley CJ, Self CH (1985) Clin. Chim. Acta 148: 119
161. Johnson HJ jr et al. (1980) J. Anal. Toxicol. 4: 86
162. Jonquey A, Touyer G: Pat. Appl. Fr. 2523975A1, 19 pp.
163. Jourdan PS, Mansell RL, Oliver DG, Weiler EW (1984) Anal. Biochem. 138: 19
164. Jourdan PS, Mansell RL, Weiler EW (1982) Planta Medica 44: 82
165. Jourdan PS, McIntosh CA, Mansell RL (1985) Plant Physiol. 77: 903
166. Jourdan PS, Weiler EW, Mansell RL (1983) J. Agric. Food Chem. 31: 1249
167. Jourdan PS, Weiler EW, Mansell RL (1985) Plant Physiol. 77: 896

168. Kado NY, Wei ET (1978) J. Natl. Cancer Inst. 61: 221
169. Kanaoka M, Yano S, Kato H, Nakano N, Kinoshita E (1983) Chem. Pharm. Bull. 31: 1866
170. Kanaoka M, Yano S, Kato H, Nakanishi K, Yoshizaki M (1984) Chem. Pharm. Bull. 32: 1461
171. Kanaoka M, Yano S, Kato H, Nakano N (1981) Chem. Pharm. Bull. 29: 1533
171a. Kanaoka M, Yano S, Kato H, Nakada T, Kawamura D (1988) Chem. Pharm. Bull. 36: 8
172. Kannangara T, Simpson GM, Rajkumar K, Murphy BD (1984) J. Chromatogr. 283: 425
173. Kato N et al. (1979) Anal. Biochem. 96: 419
173a. Kelley MM, Zahnow EW, Petersen WC, Toy ST (1985) J. Agric. Food Chem. 33: 962
174. Kemper E, Waffenschmidt S, Weiler EW, Rausch T, Schröder J (1985) Planta 163: 257
175. Khan SA, Humayun MZ, Jacob TM (1977) Anal. Biochem. 83: 632
176. Kienzle F, Mayer H, Minder RE, Thommen H (1978) Helv. Chim. Acta 61: 2616
177. Kindt TJ, Capra JD (1984) The antibody enigma. Plenum, New York
177a. Klausner A (1987) Bio/Technology 5: 551
178. Kleimola TT (1978) Br. J. Clin. Pharmac. 6: 255
178a. Knight GJ, Wylie P, Holman MS, Haddow JE (1985) Clin. Chem. 31: 118
179. Knöfel H-D, Brückner C. Kramell R, Sembdner G, Schneider K (1984) Biochem. Physiol. Pflanzen 179: 317
180. Knox JP, Galfré G (1986) Anal. Biochem. 155: 92
180a. Knox JP, Beale MH, Butcher GW, MacMillan J (1987) Planta 170: 86
180b. Kolossa E, Deus-Neumann B, Zenk MH (1987) Planta Medica 53: 449
180c. Kuniyuki AH, McCarthy S (1986) Abstr. 5C-11, 6th Intern. Congr. Pest. Chem., IUPAC, Ottawa
181. Kurogochi S, Yamaguchi I, Feyerabend M, Murofushi N, Takahashi N, Kuyama S, Weiler EW (1987) Phytochemistry 26: 2895
182. Kutney JP, Choi LSL, Worth BR (1980) Phytochemistry 9: 2083
182a. Label P, Sotta B (1988) Plant Cell Tissue Organ Cult. 13: 155
183. Lamoureux SW, Vacha WEK, Ibrahim RK (1986) Plant Science 44: 169
184. Lan HP, Gaur PK, Chu FS (1981) J. Food Safety 3: 1
185. Landsteiner K, Simms S (1923) J. Exp. Med. 38: 127
186. Landsteiner K (1945) The specificity of serological reactions. New York
187. Langone JJ, D'Onofino MR, Van Vunakis H (1979) Analyt. Biochem. 95: 214
188. Langone JJ, Van Vunakis H (1975) Res. Comm. Chem. Pathol. Pharmacol. 10: 163
189. Langone JJ, Van Vunakis H (1979) In: Albertini A, DaPrada M, Peskar BA (eds.) Radio-immunoassay of drugs and hormones in cardiovascular medicine, Elsevier, North Holland, Amsterdam, p 55
190. Langone JL, Van Vunakis H (1976) J. Natl. Cancer Inst. 56: 591
191. Lapinjoki SP, Veräjänkorwa HM, Huhtikangas AE, Lehtola TJ, Lounasmaa M (1986) J. Immunoassay 7: 113
191a. Lapinjoki S, Veräjänkorwa H, Heiskanen J, Niskanen M, Huhtikangas A, Lounasmaa M (1987) Planta Medica 53: 565
192. Lawellin DW, Grant DW, Joyce BK (1980) Appl. Environm. Microbiol. 40: 1027
192a. Lee BT, Martin P, Bangerth F (1988) J. Exp. Bot. 39: 927
193. Le Page-Degivry M-T, Bidard J-N, Rouvier E, Bulard C, Lazdunski H (1986) Proc. Natl. Acad. Sci. USA 83: 1155
194. Le Page-Degivry MT, Duval D, Bulard C, Delaage M (1984) J. Immunol. Meth. 67: 119
195. Lee S, Chu FS (1981) J. Assn. Offic. Anal. Chem. 64: 156
196. Lee SC, Chu FS (1984) J. Assn. Offic. Anal. Chem. 67: 45
197. Lee S, Chu FS (1981) J. Assn. Off. Anal. Chem. 64: 684
198. Lehmann-Kirk U, Schmidt GH, Radunz A (1979) Z. Naturforsch. 34c: 427
199. Lehtola T, Ojanen T, Huhtikangas A (1984) Farm. Tijdschr. Belg. 61: 358
200. Lehtola T, Huhtikangas A, Virtanen R (1982) Planta Medica 45: 237
201. Lenz DE et al. (1984) Fund. Appl. Toxicol. 4: 5156
202. Leroux B, Maldiney R, Miginiac E, Sossountzov L, Sotta B (1985) Planta 166: 524
203. Leute R et al. (1972) JAMA 221: 1231
204. Levitt T (1977) Lancet 2: 358
205. Levy A, Kawashima K, Spector S (1976) Life Sciences 19: 1421
206. Li PK, Lee JT, Conboy KA, Ellis EF (1986) Clin. Chem. 32: 552

206a. Li TM, Chen R, Leeder S, Stiso SN, Sizto NC, Zuk RF, Litman DJ (1987) Analyt. Biochem. 166: 276
206b. Ligler FS et al. (1987) Anal. Biochem. 163: 369
207. Litman DJ (1985) In: Ngo TT, Lenhoff HM (eds) Enzyme-mediated Immunoassay. Plenum, New York, p 155
208. Liu MT, Ram BP, Hart LP, Pestka JJ (1985) Appl. Environm. Microbiol. 50: 332
209. Loeffler LJ, Pierce JV (1973) J. Pharm. Sci. 62: 1817
210. Lowry OH, Passonneau JV (1972) A Flexible system of enzymatic analysis, Academic, New York
211. Lübke K, Nieuweboer B (1978) Immunologische Teste für niedermolekulare Wirkstoffe, Thieme, Stuttgart
212. Lukens HR, Williams CB, Levison SA, Dandliker WB, Murayama D, Baron RL (1977) Environ. Sci. Technol. 11: 292
213. Lukens HR, Williams CB, Levison SA, Dandliker WB (1977) Natl. Sci. Found., Res. Appl. Natl. Needs Report NSF/RA-770472, p 1
214. Luster MI et al. (1980) Anal. Chem. 52: 1497
215. MacDonald EMS, Akiyoshi DE, Morris RO (1981) J. Chromatogr. 214: 101
216. MacDonald EMS, Morris RO (1983) Plant Physiol. 72(S): 26
217. MacDonald EMS, Morris RO (1985) Meth. Enzymol. 110: 347
217a. Märtlbauer E, Gareis M, Terplan G (1988) Appl. Environm. Microbiol. 54: 225
218. Madej A, Häggblom P (1985) Physiol. Plant. 64: 389
219. Maggio ET (ed) (1980) Enzyme-immunoassay, CRC, Boca Raton
220. Maldiney R, Leroux B, Sabbagh I, Sotta B, Sossountzov L, Miginiac E (1986) J. Immunol. Meth. 90: 151
221. Maldiney R, Pelse F, Pilate G, Sotta B, Sossountzov L, Miginiac E (1986) Physiol. Plant. 68: 426
222. Malvano R (1982) J. Nucl. Med. All. Sci. 26: 205
223. Mansell RL, McIntosh CA, Vest SA (1983) J. Agric. Food. Chem. 31: 156
224. Mansell RL, Weiler EW (1980) In: Nagy S, Attaway JA (eds) Citrus nutrition and quality. ACS Symp. Ser. Washington, D.C., vol 143 p 341
225. Mansell RL, Weiler EW (1980) Phytochemistry 19: 1403
225a. Manz B, Grill J, Belovsky O, Kleinböhl I, Heubner A, Pollow K (1987) J. Clin. Chem. Clin. Biochem. 25: 101
226. Matsukura S, Sakamoto N, Imura H (1975) Biochem. Biophys. Res. Commun. 64: 574
227. Matthew JA, Morgan MRA, McNerney R, Chan HW-S (1983) Food Chem. Toxicol. 21: 637
228. McIntosh CA, Mansell RL (1983) J. Agric. Food Chem. 31: 319
229. Mertens R, Deus-Neumann B, Weiler EW (1983) FEBS Lett. 160: 269
230. Mertens R, Eberle J, Arnscheidt A, Ledebur A, Weiler EW (1985) Planta 166: 389
231. Mertens R, Klix D, Weiler EW (1986) 5th FESPP Congress, Book of Abstracts, Abstr. 103, Hamburg
232. Mertens R, Stüning M, Weiler EW (1982) Naturwissenschaften 69: 595
233. Mertens R, Weiler EW (1983) Planta 158: 339
234. Milstone DS, Vold BS, Glitz DG, Shutt N (1978) Nucl. Acids Res. 5: 3439
235. Moesta P, Hahn MG, Grisebach H (1983) Plant Physiology 73: 233
235a. Montillet JL, Rossignol M, Auriol P (1987) Journal of Immunoassay 8: 11
236. Morgan MRA, Bramham S, Webb AJ, Robins RJ, Rhodes MJC: Planta Medica 1985: 185
237. Morgan MRA, Coxon DT, Bramham S, Chan HW-S, Gelder WMJ van, Allison MJ (1985) J. Sci. Food Agric. 36: 282
238 Morgan MRA, Matthew JA, McNerney R, Chan HWS (1982) Proc. V. Int. IUPAC Symp. Mycotox. Phytotox., Vienna, p 32
239. Morgan MRA, McNerney R, Chan HWS (1983) J. Assoc. Off. Anal. Chem. 66: 1481
240. Morgan MRA, McNerney R, Chan HWS (1985) In: Morris BA, Clifford MN (eds) Immunoassays in Food Analysis. Elsevier Applied Science, London, p 159
241. Morgan MRA, McNerney R, Coxon DT, Chan HW-S (1985) In: Morris BA, Clifford MN (eds) Immunoassays in food analysis. Elsevier Applied Science, London, p 187
242. Morgan MRA, McNerney R, Matthew JA, Coxon DT, Chan HW-S (1983) J. Sci. Food Agric. 34: 593

243. Morris BA, Clifford MN (eds) (1985) Immunoassay in Food Analysis, Elsevier Applied Science, London
244. Morris JW, Lavender DP, Zaerr JB (1983) Plant Physiol. 72(S): 26
245. Morris RO, Akiyoshi DE, MacDonald EMS, Morris JW, Regier DA, Zaerr JB (1982) In: Wareing PF (ed) Plant Growth Substances 1982. Academic, London, p 175
245a. Morris PC, Weiler EW, Maddock SE, Jones MGK, Lenton JR, Bowles DJ (1988) Planta 173: 110
245b. Mumma RO, Brady JF (1986) 6th Intern. Congr. Pest. Chem., IUPAC, Ottawa
245c. Nagayama S, Kawamura O, Ohtani K, Ryu JC, Latus D, Sudheim L, Ueno Y (1988) Appl. Environm. Microbiol. 54: 1302
246. Neese AL, Soyka LF (1977) Clin. Pharmacol. Ther. 21: 633
247. Newsome WH, Shields JB (1981) J. Agric Food Chem. 29: 220
247a. Newsome WH (1986) 6th Intern. Congr. Pest. Chem., IUPAC, Ottawa, p 349
247b. Newsome WH (1986) Abstr. 5C-15, 6th Intern. Congr. Pest. Chem., IUPAC, Ottawa
247c. Newsome WH, Collins PG (1986) Abstr. 5C-14, 6th Intern. Congr. Pest. Chem., IUPAC, Ottawa
248. Newsome WH (1986) Bull. Environm. Contam. Toxicol. 36: 9
249. Newsome WH (1985) J. Agric. Food Chem. 33: 528
250. Newsome WH (1986) J. Assn. Offic. Anal. Chem. 69: 919
251. Nickel SL, Staba EJ (1977) Lloydia 40: 230
252. Nieschlag E, Kley HK, Usadel K-H (1975) In: Cameron EHD, Hillier SG, Griffiths K (eds) Steroid Immunoassay. Alpha Omega, Cardiff, p 87
253. Niewola Z, Hayward C, Symington BA, Robson RT (1985) Clin. Chim. Acta 148: 149
254. Niewola Z, Walsh ST, Davies GE (1983) Int. J. Immunopharmacol. 5: 211
255. Noguchi M, Matsuhita H, Tsujino Y (1978) Biochem. Biophys. Res. Comm. 83: 83
255a. Norman SM, Poling SM, Maier VP (1986) Abstr. 72, 192nd ACS Natl. Meetg., Anaheim
255b. Norman SM, Poling SM, Maier VP (1988) J. Agric. Food Chem. 36: 225
256. Odèn PC, Weiler EW, Schwenen L, Graebe IE (1987) Planta 171: 212
257. Oellerich M (1984) J. Clin. Chem. Clin. Biochem. 22: 895
257a. Ohnishi J, Nishikawa Y, Suzuki Y, Matsuda T, Miyamoto J (1986) Abstr. 5C-03, 6th Intern. Congr. Pest. Chem., IUPAC, Ottawa
258. Oksman-Caldentey K-M, Strauss A: Planta Medica 1986: 6
259. Oliver GC, Parker BM, Brasfield DL, Parker CW (1968) J. Clin. Investig. 47: 1035
260. Pabe BE (1981) Ther. Drug Monit. 3: 357
261. Parker CW, Badenoch-Jones J, Letham DS, Hocart CH, Hall PJ, Zhang R, Schröder H (1985) 12th Int. Conf. Plant Growth Substances, Book of Abstracts, Heidelberg, p 66
261a. Pauly JU, Bitter-Suermann D, Dose K (1988) Biol. Chem. Hoppe-Seyler 369: 487
262. Pence VC, Caruso JL (1986) 12th Int. Conf. Plant Growth Subst., Book of Abstracts, Heidelberg, p 68
262a. Pence VC, Caruso JL (1987) Phytochem. 26: 1251
263. Pengelly W, Torrey JG (1982) Bot. Gaz. 143: 195
264. Pengelly WL, Bandurski RS, Schulze A (1981) Plant Physiol. 68: 96
265. Pengelly WL, Hall JP, Schulze A, Bandurski RS (1982) Plant Physiol. 69: 1304
266. Pengelly WL, Meins F jr. (1977) Planta 136: 173
267. Pengelly WL (1986) In: Plant Growth Substances 1985, Springer, Berlin Heidelberg New York, p 35
268. Perry GA et al. (1984) Prepar. Biochem. 14: 431
269. Pestka JJ, Beery JT, Chu FS (1983) Food Chem. Toxicol. 21: 41
271. Pestka JJ, Chu FS (1984) Appl. Environ. Microbiol. 47: 472
272. Pesta JJ, Gaur PK, Chu FS (1980) Appl. Environm. Microbiol. 40: 1027
273. Pestka JJ, Lee SC, Lan HP, Chu FS (1981) J. Am. Oil Chem. Soc. 58A: 940
274. Pestka JJ, Łi Y, Harder WO, Chu FS (1981) J. Assn. Offic. Anal. Chem. 64: 294
275. Pestka JJ, Li YK, Chu FS (1982) Appl. Environm. Microbiol. 44: 1159
276. Pestka JJ, Lin, MT, Knudson BK, Hogberg MG (1985) J. Food Prot. 48: 953
277. Pestka JJ, Steinert BW, Chu FS (1981) Appl. Environm. Microbiol. 41: 1472
277a. Pestka JJ, Chu FS (1984) J. Food Prot. 47: 305
278. Peters H, Dierich MP, Dose KD (1982) Hoppe Seyler's Z. Physiol. Chem. 363: 1437

279. Peters JH, Baumgarten H, Schulze M (1985) Monoklonale Antikörper. Herstellung und Charakterisierung. Springer, Berlin Heidelberg New York
280. Petersen PH, Dreyer T (1976) Scand. J. Lab. Invest. 36: 103
281. Pincus SH, Watson WA, Harris S, Ewing LP, Stocks CJ, Rollins DE (1984) Life Sciences 35: 433
282. Poivier MC, Connor RJ (1982) Meth. Enzymol. 84: 607
283. Pontarotti PA, Rahmani R, Martin M, Barbet J (1985) Molecular Immunology 22: 277
284. Porcheron P, Foucrier J, Gros C, Pradeller P, Cassier P, Dray F (1976) FEBS Lett. 61: 159
285. Quarrie SA, Galfré G, Wang TL, Whitford PN (1985) 12th Int. Conf. Plant Growth Substances, Book of Abstracts, Heidelberg, p 68
286. Quarrie SA, Galfré G (1985) Anal. Biochem. 151: 389
286a. Quarrie SA, Whitford PN, Appleford NEJ, Wang TC, Cook SK, Henson IE, Loveys BR (1988) Planta 173: 330
287. Quattrone AJ, Joseph R, Jones T, Samir S, Selvaggi A, Maloney K, Lockwood T (1984) J. Med. Technol. 1: 381
288. Raikhel NV, Hughes DW, Galau GA (1987) In: Fox JE, Jacobs M (eds) Molecular biology of plant growth control, Alan R. Liss, New York, p 197
289. Rajagopal R, Ulvskov P, Marcussen J (1986) 5th FESPP Congress, Hamburg, Book of Abstracts, Abstract 16: 12
289a. Rajasekaran K, Hein MB, Davis GC, Carnes MG, Vasil IK (1987) J. Plant Physiol. 130: 13
290. Ram BP, Hart LP, Shotwell OL, Pestka JJ (1986) J. Assn. Offic. Anal. Chem. 69: 904
291. Ratcliffe WA, Fletcher SM, Moffat AC, Ratcliffe JG, Harland WA, Levitt TE (1977) Clin. Chem. 23: 169
292. Reading CL (1982) J. Immunol. Meth. 53: 261
293. Reeve DR, Crozier A (1980) In: MacMillan T (ed) Hormonal regulation of development I. Encyclopedia of plant physiology, Springer, Berlin Heidelberg New York, vol 9 p 203
294. Regier DA, Morris RO (1982) Biochem. Biophys. Res. Commun. 104: 1560
295. Rembaum A, Dreyer WJ (1980) Science 208: 364
296. Riceberg LJ, Van Vunakis H (1975) Biochem. Pharmacol. 24: 259
297. Rinder DF, Fleeker JR (1981) Bull. Environm. Contam. Toxicol. 26: 375
298. Robins RJ, Morgan MRA, Rhodes MJC, Furze JM (1984) Anal. Biochem. 136: 145
299. Robbins RJ, Morgan MRA, Rhodes MJC, Furze JM (1984) Phytochemistry 23: 1119
300. Robins RJ, Webb, AJ, Rhodes MJC, Payne J, Morgan MRA: Planta Medica 1984: 205
300a. Robotti KM, Sharp JK, Ehrmann PR, Brown LJ, Hermann BW (1986) Abstr. 42, 192nd ACS Natl. Meetg., Anaheim
301. Rodgers RM (1980) In: Maggio ET (ed) Enzyme-immunoassay. CRC, Boca Raton, p 155
302. Root MA, Gerzon K, Dyke RW (1975) In: Fed. Analyt. Chem. Spectrosc. Soc., 2nd Natl. Meetg., Abstr. 183, Indianapolis, p 125
303. Rosenthaler J, Münzer H, Voges R, Andres H, Gull P, Bolliger G (1984) Int. J. Nucl. Med. Biol. 11: 85
304. Rosher PH, Jones HG, Hedden P (1985) Planta 165: 91
304a. Ross GS, Elder PA, McWha JA, Pearce D, Pharis RD (1987) Plant Physiol 85: 46
305. Rush MD, Kutchan TM, Coscia CJ (1985) Plant Cell Reports 4: 237
306. Sandberg G, Ljung K, Alm P (1985) Phytochemistry 24: 1439
307. Sankawa U, Sung CK, Han BH, Akiyama T, Kawashima K (1982) Chem. Pharm. Bull. 30: 1907
308. Sarkar M, Mahato SB (1984) Anal. Biochem. 136: 168
309. Sayavedra-Soto LA, Trione EJ, Durley RC (1983) Plant Physiol. 72(S): 114
310. Schall RF jr., Tenoso HJ (1980) In: Immunoassays. Clinical Laboratory Techniques for the 1980s. Alan R. Liss, New York, p 127
310a. Schmidt P, Kühlmann R, Lösch U (1988) Z. Naturforsch. 43c: 167
311. Schröder G, Waffenschmidt S, Weiler EW, Schröder J (1984) Eur. J. Biochem. 138: 387
311a. Schulte-Altedorneburg M, Schneider AW, Willenbrink J (1987) J. Plant Physiol. 128: 185
312. Schwalbe M, Dorn E, Beyermann K (1984) J. Agric. Food Chem. 32: 734
313. Seitz WR (1984) Clin. Biochem. 17: 120
314. Sethi VS, Burton SS, Jackson DV (1980) Cancer Chemother. Pharmacol. 4: 183
315. Shen-Miller J, Knegt E, Vermeer E, Bruinsma J (1982) Z. Pflanzenphysiol. 108: 289
316. Shimada K et al. (1984) Chem. Pharm. Bull. 32: 2301

316a. Shirley JG, Miyamoto T, Goodrow MH, Buster D, Hammock BD (1988) J. Agric. Food Chem. 36: 863
316b. Sidki AM, Al-Abdulla IH, Rowell FJ (1987) Clin. Chem. 33: 463
316c. Singh P, Jang L (1987) Int. J. Food Microbiol. 5: 73
317. Singh BN, Galson E, Dashek W, Walton DC (1979) Planta 146: 135
317a. Sircar SM, Sircar PK, Nagar PK, Iyer RI, Chattopadhyay K (1980) Plant Biochem. J. 7: 83
317b. Skene DS, Browning G, Jones HG (1987) Planta 172: 192
318. Smith TA (1985) Annu. Rev. Plant Physiol. 36: 117
319. Smith TW, Butler VP, Haber E (1970) Biochemistry 9: 331
320. Soini E, Hemmilä I (1979) Clin. Chem. 25: 353
321. Sossountzov L, Sotta B, Maldiney R, Sabbagh I, Miginiac E (1986) Planta 168: 471
321a. Sossountzov L, Maldiney R, Sotta B, Sabbagh I, Habricot Y, Bonnet M, Miginiac E (1988) Planta 175: 291
322. Sotta B et al. (1985) J. Histochem. Cytochem. 33: 201
322a. Sotta B, Pilate G, Pelese F, Sabbagh I, Bonnet M, Maldiney R (1987) Plant Physiol. 84: 571
323. Southwick SM, Davenport TL, Castro AL (1984) Proc. 11th. Ann. PGRSA Meeting, Boston 1984
324. Spector S, Parker CW (1970) Science 168: 1347
325. Spector S (1982) Meth. Enzymol. 84: 551
325a. Spencer TM, Kitto SL (1988) Plant Cell Reports 7: 352
326. Stanker LH, Vanderlaan M, Juarez-Salinas J (1985) J. Immunol. Meth. 76: 157
327. Stanley CJ et al. (1985) Int. Clin. Prod. Rev. July/August p 44
328. Sternberger LA, Hardy PH, Cuculis JJ, Meyer MG (1970) J. Histochem. Cytochem. 18: 315
329. Sundberg B, Sandberg G, Crozier A (1986) Phytochemistry 25: 295
329a. Sundberg B, Little CHA, Riding RT, Sandberg G (1987) Physiol. Plant. 71: 163
330. Takahashi N (ed) (1986) Chemistry of Plant Hormones, CRC, Boca Raton
331. Takahashi N, Yamaguchi I, Yamane H (1986) In: Takashi N (ed) Chemistry of Plant Hormones, CRC, Boca Raton, p 57
332. Tanahashi T; Nagakura N, Inouye H, Zenk MH: Phytochemistry 23: 1917
333. Taunton Rigby A, Sher SE, Kelley PR (1973) Science 181: 165
334. Taylor HF, Burden RS (1970) Nature 227: 302
335. Teale JD, Clough JM, Marks V (1977) Br. J. Clin. Pharmac. 4: 169
336. Thong B., Soldin SJ, Lingwood CA (1985) Clin. Chem. 31: 1625
336a. Thouvenot D, Morfin RF (1983) Appl. Environ. Microbiol. 45: 16
337. Tijssen P (1985) Practice and theory of enzyme immunoassays. Elsevier, Amsterdam
338. Toshikawa T, Furuya T: Planta Medica 1985: 110
339. Treimer JF, Zenk MH (1978) Phytochemistry 17: 227
340. Trione EJ, Krygier BB, Banowetz GM, Kathrein JM (1985) J. Plant Growth Regul. 4: 101
340a. Trione EJ, Banowetz GM, Krygier BB, Kathrein JM, Sayavedra-Soto L (1987) Analyt. Biochem. 162: 301
341. Trione EJ, Morris RO (1983) Plant Physiol. 72(S): 114
341a. Tuberosa R, Sanguineti MC, Quarrie SA (1987) Giornale Botanico Italiano 121: 56
342. Turbull CGN, Hanke DE (1985) Planta 165: 366
342a. Ulvskov P, Marcussen J, Rajagopal R, Prinsen E, Rüdelsheim P, Van Onckelen H (1987) Plant Cell Physiol. 28: 937
343. Vallejo RP, Bogus ER, Mumma RO (1982) J. Agric. Food Chem. 30: 572
344. Vallejo RP, Ercegovich CD (1979) In: Trace organic analysis: a new frontier in analytical chemistry. Natl. Bureau Stand. (USA) Spec. Publ. 519, p 333
344a. Vanderlaan M, Van Emon J, Watkins B, Stanker L (1986) 6th Intern. Congr. Pest. Chem., IUPAC, Ottawa, p 597
344b. Vanderlaan M, Watkins B, Devivar R, Stanker L (1986) Abstr. 5C-02, 6th Intern. Congr. Pest. Chem., IUPAC, Ottawa
345. Van Emon J, Hammock B, Seiber JN (1986) Anal. Chem. 58: 1866
346. Van Emon JM, Seiber JN, Hammock BD (1985) ACS Sympos. Ser. 276: 307
346a. Van Emon JM, Seiber J, Hammock B (1987) Bull. Environ. Contam. Toxicol. 39: 490

346b. Van Emon JM, Seiber JN, Hammock BD (1986) Abstr. 5C-01, 6th Intern. Congr. Pest. Chem., IUPAC, Ottawa

346c. Vankova R, Kaminek M, Eder J, Vanek T (1987) J. Plant Growth Regul. 6: 147

347. Van Onckelen HA, Rüdelsheim PL, Messens E, De Greef JA (1983) Arch. Int. Physiol. Biochem. 91: B45

348. Van Peteghem CH, Van Haver GM (1986) Analyt. Chim. Acta 182: 293

349. Van Vunakis H, Farrow JT, Gijka HB, Levine L (1971) Proc. Natl. Acad. Sci. USA 68: 1483

350. Van Weemen BK, Schuurs AHWM (1971) FEBS Lett. 15: 232

350a. Völckers P, Wild A (1988) J. Plant Physiol. 133: 320

351. Vogel E, Luckner M (1981) Planta Medica 41: 161

352. Vold BS, Leonard NJ (1981) Plant Physiol. 67: 401

353. Von Schulthess GK et al. (1980) Molec. Immunol. 17: 81

354. Vonk CR, Davelaar E, Ribot SA (1986) Plant Growth. Regul. 4: 65

354a. Walker-Simmons M (1987) Plant Physiol. 84: 61

355. Walter B, Greenquist AC, Howard WE (1983) Anal. Chem. 55: 873

356. Walton D, Dashek W, Galson E (1979) Planta 146: 139

356a. Wang TL, Cook SK, Knox JP (1987) Phytochemistry 26: 2447

357. Wang TL, Futers TS, McGeary F, Cove DJ (1984) In: Crozier A, Hillman JR (eds) The Biosynthesis and Metabolism of Plant Hormones. Cambridge University Press, Cambridge, p 135

358. Warwick MJ, Bates ML, Shearer G (1985) In: Morris BA, Clifford MN (eds) Immunoassays in Food Analysis. Elsevier Applied Science, London, p 169

359. Wei R-D, Chu FS (1987) Analyt. Biochem. 160: 399

360. Weigel U, Horn W, Hock B (1984) Physiol. Plant. 61: 422

360a. Weiler EW, Eberle J, Mertens R (1986) In: Bopp M (ed) Plant Growth Substances 1985, Springer, Berlin Heidelberg New York, p 22

361. Weiler EW, Eberle J, Mertens R, Atzorn R, Feyerabend M, Jourdan PS, Arnscheidt A, Wieczorek U (1986) In: Wang TL (ed) Immunology in Plant Science. Cambridge University Press, Cambridge, p 27

362. deleted in proof

363. Weiler EW, Jourdan PS, Conrad W (1981) Planta 153: 561

364. Weiler EW, Krüger H, Zenk MH (1980) Planta Medica 39: 112

365. Weiler EW, Lach H-J (1980) Clin. Chim. Acta 104: 337

366. Weiler EW, Mansell RL (1980) J. Agric. Food Chem. 28: 543

367. Weiler EW, Schnabl H, Hornberg C (1982) Planta 154: 24

368. Weiler EW, Spanier K (1981) Planta 153: 326

369. Weiler EW, Stöckigt J, Zenk MH (1981) Phytochemistry 20: 2009

370. Weiler EW, Westekemper P (1979) Planta Medica 35: 316

371. Weiler EW, Wieczorek U (1981) Planta 152: 159

372. Weiler EW, Zenk MH (1976) Phytochemistry 15: 1537

372a. Weiler EW, Zenk MH (1981) Meth. Enzymol. 73: 394

373. Weiler EW, Ziegler H (1981) Planta 152: 168

374. Weiler EW (1984) Annu. Rev. Plant Physiol. 35: 85

375. Weiler EW (1986) In: Linskens HF, Jackson JF (eds) Immunology in Plant Science. Modern Methods of Plant Analysis. Springer, Berlin Heidelberg New York, vol 4 p 1

376. Weiler EW (1977) In: Barz W, Reinhard E, Zenk MH (eds) Plant tissue and its biotechnological application. Springer, Berlin Heidelberg New York, p 266

377. Weiler EW (1978) In: Alfermann AW, Reinhard E (eds) Production of natural compounds by cell culture methods. Proc. Int. Symp. Plant Cell Culture GBFmbH, München, p 27

378. Weiler EW (1981) Naturwissenschaften 67: 377

379. Weiler EW (1982) Physiol. Plant. 54: 510

380. Weiler EW (1979) Planta 144: 255

381. Weiler EW (1980) Planta 148: 262

382. Weiler EW (1980) Planta 149: 155

383. Weiler EW (1981) Planta 153: 319

383a. Weissenberg M, Maoz A, Levy A, Wasserman RH: Planta Medica 1988: 63

384. Westekemper P, Wieczorek U, Gueritte F, Langlois N, Langlois Y, Potier P, Zenk MH (1980) Plant Cell Physiol. 26: 1147

385. Wie SI, Hammock BD (1982) J. Agric. Food Chem. 30: 949
386. Wie SI, Hammock BD (1984) J. Agric. Food Chem. 32: 1294
387. Wie SI, Sylwester AP, Wing KD, Hammock BD (1982) J. Agric. Food Chem. 30: 943
388. Wieczorek U, Nagakura N, Zenk MH (1984) Phytochem. Soc. Symp. The chemistry and biology of isoquinoline alkaloids, Abstr. 36, London
389. Wieczorek U, Nagakura S, Sund Ch, Jendrzejewski S, Zenk MH (in press)
390. Wikman-Coffelt J, Berg HW (1976) Am. J. Enol. Viticult. 27: 115
391. Wing KD, Hammock BD, Wustner DA (1978) J. Agric. Food Chem. 26: 1328
392. Wing KD, Hammock BD (1980) Calif. Agricult. 34: 34
393. Wing KD, Hammock BD (1979) Experientia 35: 1619
393a. Wink M (1987) Pharmaceutisch Weekblad Scientific Edition 9: 217
394. Wolstenholme BN, Hofman PJ, Cutting JG, Lishman AW (1985) S. Afr. Avocado Grs. Assn. Yrbk. 8: 92
395. Wood WG (1984) J. Clin. Chem. Clin. Biochem. 22: 905
396. Woodsworth ML, Latimer LJP, Janzer JJ, McLennan BD, Lee JS (1983) Biochem. Biophys. Res. Commun. 114: 791
397. Woychik NA, Hinsdill RD, Chu FS (1984) Appl. Environm. Microbiol. 48: 1096
397a. Wraith MJ, Hitchings EJ, Woodbridge AP, Cole ER, Roberts TR (1986) Abstr. 5C-10, 6th Intern. Congr. Pest. Chem., IUPAC, Ottawa
398. Wyndaele R, Van Onckelen H, Christiansen J, Rüdelsheim P, Hermans R, De Greef J (1985) Plant Cell Physiol. 26: 1147
398a. Xu Y-C, Zhang G-S, Chu FS (1986) J. Assn. Offic. Anal. Chem. 69: 967
399. Yamaguchi I, Nakagawa R, Kurogochi S, Murofushi N, Takahashi N, Weiler EW (1987) Plant Cell Physiol. 28: 815
399a. Yoshioka M, Aso C, Amano J, Tamura Z, Sugi M, Kuroda N (1987) Biogenic Amines 4: 229
399b. Yoshioka M, Iwai-Yamasaki M, Kurakazu-Wada K, Shirahata A, Tamura Z (1987) Biogenic Amines 4: 211
219
400. Zaerr JB, Durley RC, Morris RO (1983) Plant Physiol. 72(S): 26
401. Zavala ME, Brandon DL (1983) J. Cell Biol. 97: 1235
402. Zhang GS, Schubring SL, Chu FS (1986) Appl. Environm. Microbiol. 51: 132
402a. Zhang GS, Schubring SL, Chu FS (1986) Appl. Environm. Microbiol. 51: 132
402a. Zhang J, Davies WJ (1987) J. Expt. Bot. 38: 649
403. Zuk RF et al. (1985) Clin. Chem. 31: 1144

Subject Index

Vol. 1

Sterol Biosynthesis Inhibitors and Anti-Feeding Compounds

Sterol-Biosynthes in Fungi, a Target for Broad Spectrum Fungicides
T. Kato

Chemistry of Sterol Biosynthesis Inhibiting Fungicides
D. W. Krämer

Biology of Sterol-Biosynthesis Inhibiting Fungicides
K.-H. Kuck, H. Scheinpflug

Anti-Feeding Compounds
D. M. Norris

Vol. 2

Degradation of Pesticides, Desiccation and Defoliation, ACh-Receptors as Targets

Microbial Degradation of Pesticides
P. R. Wallnöfer, G. Engelhardt

Enzymes Involved in the Metabolism of Organophosphorus, Carbamate and Pyrethroid Insecticides
U. Schoknecht, D. Otto

Insecticide Compounds, Acting on the Acetylcholine Receptor of the Insect Nervous System
R. Richter, D. Otto, H.-J. Mengs

Chemical Manipulation of Desiccation and Defoliation and Essential Aspects for the Application and Development of New Chemical Compounds in the Future
H. Bergmann, D. Martin

MIX
Papier aus verantwortungsvollen Quellen
Paper from responsible sources
FSC® C105338

If you have any concerns about our products,
you can contact us on
ProductSafety@springernature.com

In case Publisher is established outside the EU,
the EU authorized representative is:
Springer Nature Customer Service Center GmbH
Europaplatz 3, 69115 Heidelberg, Germany

Printed by Libri Plureos GmbH
in Hamburg, Germany